"Her book (thankfully) is more like an essay than like a treatise. Heffernan is smart, her writing has flair, she can refer intelligently to Barthes, Derrida, and Benjamin—also to Aquinas, Dante, and Proust—and she knows a lot about the Internet and its history."

—*New Yorker*

"This is sumptuous writing, saturated with observations that are simultaneously personal, cultural, and strikingly original—and she's writing about software. I love it. Ultimately, the art here is her prose style."

—*New Republic*

"One of the writers I most admire."

—Gwyneth Paltrow

"Marrying this study with her own fascinating personal history with the internet as a pre-teen, *Magic and Loss* is a revealing look at how the internet continues to reshape our lives emotionally, visually and culturally."

—*Smithsonian*

"The best writing on Angry Birds you'll ever encounter."

—*Wired*, #1 Summer Beach Read

"Heffernan's rhetoric is so dexterous that even digital pessimists like me can groove to her descriptions of 'achingly beautiful apps,' her comparison of MP3 compression to 'Zeuxis's realist paintings from the fifth century BC.' And Heffernan is subtly less optimistic than she at first seems—she knows that magic is not the opposite of loss, but sometimes its handmaiden. She's written a blazing and finally wise book, passionate in its resistance to the lazy certitudes of a cynically triumphal scientism."

—Michael Robbins, author of *The Second Sex* and *Alien vs. Predator*

MAGIC

AND

LOSS

THE INTERNET AS ART

VIRGINIA HEFFERNAN

SIMON & SCHUSTER PAPERBACKS

NEW YORK   LONDON   TORONTO   SYDNEY   NEW DELHI

For Ben and Susannah

Simon & Schuster Paperbacks
An Imprint of Simon & Schuster, Inc.
1230 Avenue of the Americas
New York, NY 10020

First Simon & Schuster trade paperback edition June 2017

SIMON & SCHUSTER PAPERBACKS and colophon are registered
trademarks of Simon & Schuster, Inc.

For information about special discounts for bulk purchases,
please contact Simon & Schuster Special Sales at 1-866-506-1949
or business@simonandschuster.com.

The Simon & Schuster Speakers Bureau can bring authors
to your live event. For more information or to book an event,
contact the Simon & Schuster Speakers Bureau at 1-866-248-3049
or visit our website at www.simonspeakers.com.

Interior design by Ruth Lee-Mui

Manufactured in the United States of America

10 9 8 7 6 5 4 3 2 1

Library of Congress Cataloging-in-Publication Data is available.

ISBN 978-1-4391-9170-5
ISBN 978-1-5011-3267-4 (pbk)
ISBN 978-1-4391-9171-2 (ebook)

Parts of this book were published, in different forms,
in the *New York Times*, Yahoo! News, and *Wired*.

CONTENTS

PREFACE

On January 25, 2006, a mysterious image showed up on You-Tube, the video-sharing site that was then only three months old. A sinewy figure in a swimming-pool-blue T-shirt, his eyes obscured by a beige baseball cap, was playing electric guitar. Sun poured through a window behind him. He played in a yellow haze. The video was called simply *guitar*. A handmade title card gave the performer's name as Funtwo.

The piece Funtwo played with mounting dexterity was an exceedingly difficult rock arrangement of Pachelbel's Canon, the composition from the turn of the eighteenth century known for solemn chord progressions and overexposure at weddings. But this arrangement, attributed on another title card to someone called JerryC, was anything but plodding: it required high-level mastery of a singularly demanding maneuver called sweep-picking.

Over and over the guitarist's left hand articulated strings with barely perceptible movements, sounding and muting notes almost simultaneously, and playing complete arpeggios with a single stroke of his right hand. The video was thrilling to watch.

Almost instantly I was hooked. I hadn't yet seen selfies of any kind, handheld or selfie-stick-enabled, nor had I seen video on Skype or FaceTime, so I wasn't accustomed to this intensely focused exhibitionism, the pleasingly distorted self-portraits in moving pixels, often of family and intimate friends, that now flood our screens. Funtwo's own selfie video was curious, masturbatory: David Hockney colors plus chiaroscuro. The effect was not wholesome. The video lacked the creamy resolution, crystalline audio, and voluptuous effects associated with professionalism—and with even the average MTV entry.

Amateur. Homemade. Flawed. Not so much mesmerizing as provocative. Harold Bloom wrote that to behold is a tragic posture; to observe is an ethical one. Funtwo required near-clinical observation. You didn't *behold* this video, as you might a Hollywood movie, enraptured by the spectacle. You inclined toward it. You studied it, like a scientist. You peered, as at scrambled porn on a high and forbidden channel.

As soon as I leaned forward, I had reached for Tolkien's ring, or tasted some life-altering drug, or crossed a magical line, and there was no going back. Just as Nabokov forces us to take Humbert Humbert's language into our very mouths in the opening of his great novel of child rape—"Lo-lee-ta: the tip of the tongue taking a trip of three steps down the palate to tap, at three, on the teeth"—this video seemed to implicate anyone who watched it.

I played *guitar* again, then again. A small miracle was quietly happening in those first months on the site. The bona fide pornography that was widely expected to drive out all other video genres—as a predator plant strangles diverse flora and unbalances ecosystems—never showed up. Without actual porn, the subtler voyeurism of *guitar* stood a chance of becoming a hit with viewers. And hit it was. By the end of its first week on YouTube the video had been viewed 1 million times. By 2016 its various versions had drawn more than 10 million views, and for years it was regularly listed among the most-seen snippets of online video in the history of the World Wide Web.

Working as a critic and columnist for the *New York Times*, I had acquired some unusual new habits since YouTube launched. *Guitar* only threw the problem into relief. As network television contracted, the media business folded dozens of magazines, and YouTube was acquired by Google for $1.65 billion, with other $1 billion–plus tech acquisitions and giant IPOs in the offing, I found myself mystified by how much time I spent away from the tattered-armchair totems of my youth: books, magazines, newspapers, the broadcast networks, and the ever-present murmur of NPR.

While there was still achievement and pleasure in the old media, it was clear too that the dogs had barked; the great caravan that brings the knowledge and ideas that shore up human enterprises had moved on. I renewed my subscriptions to *Vogue* and the *New York Review of Books*, until I didn't anymore. Back issues had piled up on my coffee table and then become part of recycling, landfills, and compost. They weren't culture; they were carbon. Part of the problem gumming up the environmental works.

The same thing happened to the novels—Hilary Mantel, a reissue of John Updike—that I ordered in hard copy from Amazon. The spell that had been cast over me by inked letters on white pulp was broken. Or more accurately: a new spell had been cast, on a separate part of my brain.

The deeper I ventured into the civilization I found online, the more I realized I'd need new models of courage and imagination to contend with the trippy, slanted, infinite dreamland of the rapidly evolving Web. Funtwo became my hero. The velocity, intricacy, and exactness of his performance modeled the rhythms and mental requirements of the Web itself.

Funtwo's *guitar* video speaks to me now, a decade later, just as a chalice of certain dimensions tells us something about the people who inhabited a lost world. From a chalice we learn how big were the hands that were meant to hold it; how much liquid people liked at once and could consume; what kind of liquid, cold or hot, basic or acidic, they considered potable; what type of surface their cups might sit on. The dozens of hours that I spent feeding my obsession with *guitar* were not wasted. Or not entirely.

The video was, in fact—as Funtwo (né Jeong-Hyun Lim) told me when I finally met him—intended to be instructional, an early contribution to the now encyclopedic how-to category on YouTube. (Which I would later consult to learn umbilical care for my infant daughter, as well as how to shield an iPad from scratches and how to make progress in a Wii game called Lego City Undercover.) For me the video contained a powerful suggestion of the kind of person I would have to become if I was to keep a clear head in the new medium that had come to dominate my mental life. And I was not alone: the Internet was pervading the lives of all of us who were growing into a newly transmogrified social and aesthetic space, from my neighbors and colleagues, friends and children, to musicians in Taiwan and Seoul and all of the 1.4 billion active users of Facebook.

In the Funtwo days, well before the efflorescence of elegant services like Spotify for music, Reddit for ideas, Pinterest for collages, and Instagram for photographs; before Steve Jobs's death; and even before the iPhone, the socialization and mobilification of everything, and then the move to wearables, the so-called Internet of Things, 3D printing, and virtual reality, the Web asserted itself as its own culture. Right at the dawn of Web 2.0, when newly expansive broadband permitted the dissemination of video and the rise of social networking, the Internet became something more than a reformulation of the offline world. With cries variously of agony and triumph, we had to stop pretending that email was a handy alternative to telephones or post. Fluid and never-ending electronic exchanges made the word *communication* seem inadequate. Similarly newspapers on the Web could no longer be considered mere adaptations of newsprint.

1992 FLASHBACK

A traumatic moment in early adulthood came just as I was having doubts about my first choice of career: academia. I had started a PhD program in English in 1991, just after graduating from college. But though I was supposed to be refining my skills at rigorous academic prose, the grubbier work of Greil Marcus, whose book *Lipstick Traces* took seriously pop culture, in a pop idiom, captured my attention. I also sometimes tried in vain to copy Camille Paglia's outlandish and supersexual observations about art, though they were considered suspect in university settings. One day my traditionalist father nailed me for this.

Dad: Virginia, your prose can be a touch glib—or, rather, meretricious.

Me: W-what's meretricious, Dad?

Dad: Oh Virginia! Come on!! MERETRIX. From your Latin. "Like a prostitute."

Me: Oh—I—Oh.

Dad: Did you get NOTHING out of Latin Camp?

My father called me a prostitute? This is not an easy dialogue to recall, but eccentricities of the early 1990s are a useful reference point when taking the measure of the Internet's influence. Those were the days before the Web. The Mount Vesuvius of digitization was faintly rumbling, but most of us were determined to block out the noise. Sure, there was email, but texting and tweeting had not yet made glibness compulsory. The Meretrix, by other names, had not yet become an Instagram paragon. Mandarin and emojis had not yet left Latin in the dust as second languages of choice. These were exciting times, filled increasingly with desktop-published zines and other transitional forms that presaged blogs, but cultural loyalists were still hoping to hold on to old paradigms as long as possible.

Today holding on is just about impossible. The tectonic shift has happened. The *New York Times* daily newspaper and the company's news apps are starkly disparate entities, and only one of them is defined by short-form aggregation, data visualizations, and streamable video. Uber is something other than a municipal taxi service. Airbnb is not another kind of hotel. Ecommerce—at eBay, Amazon, Etsy—is not analogous to catalogue shopping; it has its own rules, conventions, implications, pace, and prices. Between analog and digital are more than differences in degree. Between them is a difference in kind.

Like all new technologies, the Internet appears to represent the world more faithfully than the technologies that preceded it. And the Internet is an *extraordinarily* seductive representation of the world. We've never seen a work of art like it. That is this book's central contention: that the Internet is a massive and collaborative work of realist art. Moreover it's so beguiling a realist showpiece, and so readily confused with reality, that books about it call themselves books about "business," "politics," or "science"—the reigning bywords for reality. That's a mistake. Digital forms are best illuminated by cultural criticism, which uses the tools of art and literary theory to make sense of the Internet's glorious illusion: that the Internet is life.

Because of course the Internet is not life. In fact it is a highly artificial regime, with tight rules and rituals that organize its text, music, and images. That's why the Internet becomes more deeply meaningful and moving when "read" as an aesthetic object than lived or reported on as firsthand human experience. That human experience is art, where art is considered closer to a game than to a deception. Our proxies in this game are our avatars: the sum total of all the profile pictures, message-board communiqués, Snapchat videos, and all other artifacts of text, image, and sound that we add to the Internet and attach to our various handles. The game itself, an artwork, is without doubt what video gamers call an MMORPG: a massively multiplayer online role-playing game.

Digital life, in its current extremely visual, social, portable, and global incarnation, rewards certain virtues. They're not the ones many of us grew up with. Engagement, emotional expression, liberalism, tolerance, self-knowledge, irony: these values of the 1970s, refined while I was in college and then in graduate school in the 1990s, lost a great deal of urgency after the turn of the millennium. It was unnerving to watch them go. How long had we all dilated on and argued over whether a poem ought to be read as an independent artifact or in historical context, whether a professor ought ever be fired for her views on Israel or date rape, and whether a scientific or cultural worldview was more accurate. It went on and on: what everyone made of Monica Lewinsky, Yasser Arafat, Reaganomics, the semiotics of hip-hop, the cold war, or the implications of the Milgram experiment. We citizens used all the language and logic at our command to parse these problems, with our government and institutions mostly emphasizing liberalism and tolerance. But now that digitization has changed even knowledge and ethics, the values instilled in me as the daughter of a Latin-besotted college professor in New England have turned slightly old-fashioned, like the notion of fame in *Beowulf* or honor in Sir Walter Scott's novels.

What I was trying to learn as I practiced the finger work required by my laptop, BlackBerry, and eventually iPhone—and what writers, workers, teachers, parents, students, artists, and companies appeared to be trying to learn too—were new skills and interpretive methods, many of which didn't have names yet.

After dusting off hundreds and then thousands of videos on YouTube, I have begun to see clearly the civilization they compose. Online video isn't a *new* art form, I discovered, like punk music or color-field painting in their time, starting in a time and place and slowly burgeoning. Instead the art of the Internet and its rules came into view all at once and fully formed. All over the world amateurs had apparently spent the years since the birth of camcorders (in 1982) and digital video (in 1986) shooting, producing, collecting, or transferring home movies, video art, pet and baby videos, surveillance videos (including some that showed police, interpersonal, and corporate misdeeds), music performances, ads, trailers, sermons, lectures, comedy sketches, theatrical scenes, pornography, magic tricks, athletic stunts, pranks, virtual tours, news broadcasts, video op-eds, how-to videos, and a vast reserve of unclassifiable entries that needed only an audience.

By 2016 more than one hundred hours of video were uploaded every minute to YouTube—hours that came with a dizzying range of styles, themes, and provenances. Many of them had clearly been produced well before the possibility of online broadcasting even existed. The first videos mounted to YouTube included a scene of civil disobedience shot in a bus in Singapore, a monologue by a Best Buy clerk, and fully fourteen short movies of mammals playing with shoelaces. Every single one of them zinged around the Web and, collectively, attracted far more passionate responses than the multimillion-dollar slates of new network television shows I regularly reviewed. For anyone (from college kids to CEOs) trying to understand our speedy, freshly digital world the videos were and are invaluable. They show whole new facets of human experience.

The Internet favors speed, accuracy, wit, prolificacy, and versatility. But it also favors integrity, mindfulness, and wise action. For however alien in appearance, the Internet is a cultural object visibly on a continuum with all the cultural artifacts that preceded it. It is not a break with history; neither is it "progress." It's just what happened to be next. It is not outside human civilization; it is a new and formidable iteration of that civilization. It's also a brilliant commentary on it. To be still more specific: the Internet responds, often with great sensitivity, to critical methodologies. Sense can be made of it. Logic can be divined in it. Politics can be derived from it. Pleasure can be taken in it. Beauty can be found in it. Pain too—and loss. Agony and ecstasy is what I mean: the Internet may not be reality, but it's very real art.

This has become plain in the development of hundreds of new discourses online, including feminist hawkism on FrontPageMag .com ("Inside every liberal is a totalitarian screaming to get out") and French pro-Americanism on Médiapart. It surfaces in the new-media presidency of Barack Obama, which had its policies inflected and even set by the exigencies of the Internet. I saw it clearly when I tracked down wily producers of a hoax series called *lonelygirl15* and eventually even in the captivating Funtwo, the fame-averse Korean guitarist who taught me how an export-driven economy like South Korea's, which is long on cultural producers and short on cultural consumers, transforms even the way music is made and musical genres refined.

The Internet's responsiveness to critical tools—the kind used by English majors, historians, bloggers, readers of every stripe, including rogues like commenters, trolls, and knee-jerk tweeters—has been elucidated in my studies of baroque audio-visual projects by composers and sound designers who get it, including the Israeli Kutiman, the American Beyoncé, and the Swede Paul N. J. Ottosson.

The hallmarks of Internet culture come through in experiments like Netflix and Amazon originals, arguments and reports serialized on Twitter, podcasts like NPR's *Serial*, and the new-media franchises of reality-TV heroines. Anyone can witness from the front row the emergence of a new hierarchy of values at Wikipedia, Facebook, Twitter, Instagram, Spotify, Snapchat, Skype, Yahoo!, Tumblr, Quora, eBay, Amazon, Seamless, game apps, YouTube, the Kindle, the iPhone, the iPad, FitBit, Google Glass (RIP), Oculus Rift, the Amazon Echo, message boards, and a world of blogs and commentary—through their rises and rises and rises. And, in several cases, their fascinating falls.

But the companies rise and fall on the strength and value to advertisers of what at Yahoo! News, where I covered digital politics during the 2012 election, we used to optimistically call their "assets"—visual, auditory, textual. (At least they weren't liabilities.) These assets are nearly always ironic, cartoonish, or dramatic extensions of established and even ancient art forms: aphoristic poetry suitable for Twitter; painterly images for Instagram; polemics, essays, and reports for Facebook.

When the comedy-drama series *Orange Is the New Black* appeared in 2013 as one of Netflix's first batch of original dramas—"television" had long been all-digital by then—viewers took it in stride that it would be an inmate's-eye story. It was the latest expression of prison literature, preceded by works from Socrates, Jack Henry Abbott, and Nelson Mandela, among others. These prisoners wrote books while in prison because they had paper, pens, and time. Why not? Newspapers had given us the numbers, but only popular culture could fully capture the reality that Americans live in an era of widespread imprisonment. The nation is striped with a penal colony that runs coast to coast. A hundred thousand Web users could have told you that prison literature had *already* taken many new shapes in the twenty-first century and dug roots deep into the Internet—on, for example, the so-called Big Board, Prison Talk Online. This large and polyglot message board, conceived in a jail cell by a felon named David Frisk, which in its first decade attracted nearly 7 million posts, connects a vast network of prisons around the world. Families check it for real-time word of prison fights. Prisoners post poetry on it. Legal advice is given. Threads range from "How to Lose 50+ lbs. before Your Man Comes Home" to "Preparing for Executions." For readers, reporters, and concerned citizens, no document more urgently suggests the intricacies of the world's hyperextended prison system: fine-grain logistical detail and harrowing incongruities. *Orange Is the New Black*, as a literary and political artifact based on a literary memoir, can only aim to distill and dramatize the wisdom of that massive project.

Other examples make still more obvious that Internet art is not all marginalia and kitty kitsch. Like Prison Talk, the June 2009 video of the murder of Neda Agha-Soltan during a march protesting the election in Iran of President Mahmoud Ahmadinejad—a video that played in Iran and around the world—crystallized a new political reality. Thanks to ubiquitous digital cameras and instant global dissemination techniques,

Agha-Soltan became a portable symbol of the antigovernment movement in Iran the very day she died. But she's not dead in her symbolic form; she's dying, perpetually, blood pouring almost audibly from her as she lies in the street, having been shot in the heart by a sniper.

That this video circulated widely on Twitter and Facebook and instantly gained an entry in the Persian-language Wikipedia (written by the fiancé of Agha-Soltan) contributes to the effect of motion. The chaos of the protests in Iran was echoed and amplified in the chaotic Twitter reports and made the violence seem immediate and urgent even to Europeans and Americans. But when it seemed to end—when "#Neda" was no longer a trending topic on Twitter, when the video's viewers capped at around 1 million on YouTube—did the movement in Iran strike the global community as somehow finished? Unlike after events that attracted the predigital rhetoric of martyrdom (the discourses around Kent State and Ruby Ridge, say), loyalists did not demand that Agha-Soltan not be forgotten. They didn't wear her image in lockets or on T-shirts. They didn't rehearse the circumstances of her death. After the video made its rounds, it seemed we had all been eyewitnesses to the protests and the murder, and—as horrifying as it had been—it was over.

Tweets, Facebook posts, and YouTube videos seem like discrete media entities, and it's easy to focus on how illiterate they are, or how trivial. But just as the American Revolution can be seen as a consequence of the pamphlet, and the antiwar movement can be seen as rooted in television and photojournalism, the rise of various contemporary ideologies—Ron Paul libertarianism, Obama idolatry, and "fact-check" politics—can be seen as epiphenomena of the Internet.

To make sense of the new world we are living in—in all its speed, diversity, and eccentricity—to truly fathom the high-velocity and rapacious new medium that has both re-created and shattered traditional forms, we need to risk the pain and scrap our old aesthetics and consider a new aesthetics and associated morality.

A new brand of intellectual courage must be brought to envisioning this new symbolic order. For artists, ignoring the imperative to grasp the cultural implications of the Internet means risking irrelevance. For companies, devaluation. For politicians and foreign policy architects, it means incomprehension about how meaning is configured, with a resulting foundering of campaigns, administrations, and initiatives. As human discourse adapts to its new home, everything we do and think as human beings will be and is being shaped by new values.

Magic and Loss starts with the building blocks of our digital culture. In analog life these cultural blocks might be considered literature and communication, visual art, film and television, architecture, fashion and design, food, sculpture, dance, and music. Online, in pixels, where flesh, marble, and 3D space is (so far) scarce, it's somewhat simpler: design, text, photography, video, and music. My aim is to build a complete aesthetics—and poetics—of the Internet.

Any book about the Internet ought to offer a useful structure for the headache-inducing chaos of digital life. *Magic and Loss* does this, and also proposes how its pleasures might be savored— the way Ian Watt and Leslie Fiedler showed readers how to approach novels, Pauline Kael showed us how to approach movies, Lester Bangs and Greil Marcus showed us how to approach rock music, Susan Sontag showed us how to approach photography, and George Trow and Marshall McLuhan showed us how to approach media.

The Web represents a grand emotional, sensory, and intellectual adventure for anyone willing to explore it actively. Alarmist tracts that warn about how the Web endangers culture or coarsens civilization miss the point that the same was said in turn about theater, lyric poetry, the novel, film, and television. I want instead to show how readers might use the Web and not be overwhelmed by it; how we might stop fighting it, in short, and learn to love its hallucinatory splendor.

The Internet is the great masterpiece of human civilization. As an artifact it challenges the pyramid, the aqueduct, the highway, the novel, the newspaper, the nation-state, the Magna Carta, Easter Island, Stonehenge, agriculture, the feature film, the automobile, the telephone, the telegraph, the television, the Chanel suit, the airplane, the pencil, the book, the printing press, the radio, the realist painting, the abstract painting, the Pill, the washing machine, the skyscraper, the elevator, and cooked meat. As an idea it rivals monotheism.

Just as, in Nietzsche's scheme, man created science, which in turn killed god, analog culture—books, clocks, film, industrial machines, the compasses and timers of scientific method—created digital culture, and now digital culture has superseded it. It was quick, the supersession—and now it's over. But where are we?

Magic is a word that Apple vigorously embraced. The iPad was introduced as a "magical and revolutionary device." And *magic* is a crucial term of art in computer programming. Computer code is considered magic when it seems simple but accomplishes complex operations. The Internet is paradigmatic magic. It turns experiences from the material world that used to be densely physical—involving licking stamps, say, or winding clocks or driving in cars to shopping centers—into frictionless, weightless, and fantastic abstractions. As Lawrence Lessig puts it, "The digital world has more in common with the world of ideas than with the world of things."

And yet it's still here, the persistent sense of loss. The magic of the Internet—the recession of the material world in favor of a world of ideas—is not pure delight. It seems we are missing something very worthwhile and identity-forming from our pre-digital lives. Is it a handwritten letter? Is it an analog phone call? Is it a quality of celluloid film, a multivolume encyclopedia, or a leather-bound datebook? Is it a way of thinking or being or even falling in love?

Between two discourses, two languages, two regimes, something is *always* lost. And whether or not we admit it, the Internet and its artifacts are not just like their cultural precedents. They're not even a rough translation—or a strong misreading—of those precedents. The Internet has a logic, a tempo, an idiom, a color scheme, a politics, and an emotional sensibility all its own. Tentatively, avidly, or kicking and screaming, nearly 2 billion of us have taken up residence on the Internet, and we're still adjusting to it.

This transformation of everyday life includes moments of magic and an inevitable experience of profound loss. Any discussion of digital culture that merely catalogues its wonders and does not acknowledge these two central themes is propaganda and fails to do it justice.

Thirty-five years ago, when I first discovered it, the Internet wasn't easy to find. It wasn't a user-friendly retail franchise, as the Web is now. It was a nervous back office full of furtive clerics. You stumbled in. While computer hardware and software of the 1970s were the work of sophisticated engineers who pressed computers into the service of everything from music to word processing, architecture, and filmmaking, the slow and awkward networks in those days had limited application. These were the so-called eve networks, inspired largely by ARPANET, the landmark computer–communications system that was a project of the U.S. government's Advanced Research Projects Agency. Logically the Internet in its early days was a kind of diversion for cold war intelligence types and academics. But it was possible to stumble onto the early Internet.

I know because I was among the stumblers. Xcaliber was early social-networking technology developed at Dartmouth College. In the heyday of Dungeons & Dragons, its vaguely Arthurian theme appealed to both hackers and tweens. Its real purpose was to facilitate communication among the several academic and scientific institutions that shared Dartmouth's mainframe computer—one of those big, heaving rhinos in a cage of bulletproof Plexiglas. Every day a few hundred people dialed that mainframe for an alien signal—the then-unfamiliar squeal and crash of information transmission—and fit their receivers into acoustic couplers, like people in kayaks.

As a townie preteen, I hacked in with the help of some shaggy, kind Dartmouth students who called themselves sysprogs. In those days, "Dartmouth sysprog" sounded tantalizing to me—the way "lead singer" sounded to some of my classmates. John Kemeny, then the president of Dartmouth, had cowritten the computer language BASIC in 1964, inspiring a generation of student programmers to trek north to our snowy town. This group ("Kemeny's Kids") built the extraordinary Dartmouth Time-Sharing System, which allowed people from all academic departments, even the humanities, to use a computer network. The sysprogs of the 1970s and early '80s also tended the mainframe as it shook and rattled incongruously on the edge of Dartmouth's Colonial campus.

With some friends I found my way to their computer center under the pretext that we wanted to talk about BASIC. We were lucky to have this opening sally. Kemeny had been required to teach rudimentary BASIC to the local Yankee schoolchildren, presumably to win the freedom to pursue his decidedly non-Yankee plan for ARPANET-on-the-Connecticut. To mollify our parents, we told them we needed to sharpen our programming skills if we were ever going to "work for NASA."

But the little girls of ten and eleven who stormed Xcaliber never made tech history. That worked for us at the time: all the better to enter the shadowy world of Xcaliber—and especially an addictive live-chat feature called Conference XYZ—without being noticed. Conference XYZ amplified Xcaliber's fantasy element: each convocation had levels and a self-anointed master who could banish chatters he disliked. Participants often communicated in an odd Led Zeppelin idiom or referred to damsels and steeds. I loved this. Under cover of my first avatar, Athena (naturally), I learned all of the digital skills I still rely on today. I learned to type, to talk rapidly in entwined threads with several people at once, to experiment with idioms, to test and learn, to recover from reply-all mistakes, to spot lechy people by their online styles, and to avoid ideologues who post in all caps. Most important of all, I learned, as a novelist does, to create an avatar in digital space who is simultaneously flawed, dignified, and realistic—but who can also field trolls and take sniper fire for me and thus keep the *real* me, my soul, entirely aloof and safe.

By the time I turned thirteen, I was confident I knew *every single person* online. My parents couldn't have guessed I was meeting anyone. As I sat alone at the computer hour after hour it seemed I was learning "computers." In fact, I was learning culture.

The story of early computer networks has most often been told as a technology and business story. But like the Internet today, Conference XYZ was not an engineering experiment as much as an immersive experience. What mesmerized me and its other users were its cadences and its vocabulary. Its vibe. On some level, while we were seeking connection and community we were also helping to build a culture. Today I see that culture writ large online.

Conference XYZ pretty much folded in 1986. For years I half-repressed thoughts of Xcaliber. It would come to me in fragments of memories: the odd jargon we evolved, the hot feeling of being watched, the invective, the jokes, the speed. The highly collaborative project had been the spontaneous creation of a scene, a modus vivendi, an entire culture. Had we really done all that? And was it really gone?

It was not gone. What I thought was the end of a short detour from regular life was actually the beginning of the biggest cultural phenomenon of my lifetime. If it's ever fair to say that anything has "changed everything," it's fair to say so about the Internet. At stake in this cultural transformation are the way we live, the way we think, the way we love, the way we talk, and even the way we fight across the globe. The Internet is entrenched. It's time to understand it—and not as a curiosity or an entry in the annals of technology or business but as an integral part of our humanity, as the latest and most powerful extension and expression of the project of being human.

1

DESIGN

Instead of introducing a narrative or a lyric structure, an app game called Hundreds begins with a hazy dynamic: *expanding*. A player meets no characters; rather she's put in mind of broadening her horizons, dilating on a subject, swelling with pride. Cued by dreamlike graphics, she feels her neurons inflate.

Next she's abstractly navigating a crowd in that expansive state. She's flinching to keep from touching anyone else. Then, on top of all that, she is shot through with the urgent need to get someone alone, to guide him away from the crowd. Finally she's doing this while trying to avoid the blades of a low ceiling fan.

These obscure neurological half-narrative states and others, far stranger, are cunningly evinced by Hundreds, which is a masterpiece mobile puzzle game by Greg Wohlwend and Semi Secret Software. As in Hundreds (and 1010!, Monument Valley, and the marvelous blockbuster Minecraft), much of the best digital design bypasses language and can only be evoked by it, not denoted precisely.

Superb and sleek digital design like Semi Secret Software's now live on apps. These apps are not so much intuitive as indulgent, and they put users far from the madding crowd of the World Wide Web. The extreme elegance of app design has surfaced, in fact, in reaction to the extreme inelegance of the Web.

Appreciating the Web's entrenched inelegance is the key to understanding digital design both on- and offline. Cruise through the gargantuan sites—YouTube, Amazon, Yahoo!—and it's as though modernism never existed. Twentieth-century print design never existed. European and Japanese design never existed. The Web's aesthetic might be called late-stage Atlantic City or early-stage Mall of America. Eighties network television. Cacophonous palette, ad hoc everything, unbidden ads forever rampaging through one's field of vision, to be batted or tweezed away like ticks bearing Lyme disease.

THE ADMIRING BOG

Take Twitter, with its fragmentary communications and design scheme of sky-blue birdies, checkmarks, and homebrew icons for retweets, at-replies, hashtags, and hearts. It's exemplary of the graphic Web, almost *made* to be fled. Twitter's graphics can be crisp and flowy at once, if you're in the mood to appreciate them, but the whole world of Twitter can rapidly turn malarial and boggy. The me-me-me clamor of tweeters brings to mind Emily Dickinson's lines about the disgrace of fame: "How public—like a Frog—/To tell one's name—the livelong June—/To an admiring Bog!"

That boggy quality of the Web—or, in city terms, its ghetto quality—was brought forcefully to light in 2009, in a sly, fuck-you talk by Bruce Sterling, the cyberpunk writer, at South by Southwest in Austin, Texas. The Nietzschean devilishness of this remarkable speech seems to have gone unnoticed, but to a few in attendance it marked a turning point in the Internet's unqualified celebration of "connectivity" as cultural magic. In fact, Sterling made clear, connectivity might represent a grievous cultural loss.

Connectivity is nothing to be proud of, Sterling ventured. The clearest symbol of *poverty*—not canniness, not the avant-garde—is dependence on connections like social media, Skype, and WhatsApp. "Poor folk love their cell phones!" he practically sneered. Affecting princely contempt for regular people, he unsettled the room. To a crowd that typically prefers onward-and-upward news about technology, Sterling's was a sadistically successful rhetorical strategy. "Poor folk love their cell phones!" had the ring of one of those haughty but unforgettable expressions of condescension, like the Middle Eastern treasure "The dog barks; the caravan passes."

Connectivity is poverty, eh? Only the poor, defined broadly as those without better options, are obsessed with their connections. Anyone with a strong soul or a fat wallet turns his ringer off for good and cultivates private gardens (or mod loft spaces, like Hundreds) that keep the din of the Web far away. The real man of leisure savors solitude or intimacy with friends, presumably surrounded by books and film and paintings and wine and vinyl—original things that stay where they are and cannot be copied and corrupted and shot around the globe with a few clicks of a keyboard.

Sterling's idea stings. The connections that feel like wealth to many of us—call us the impoverished, we who brave Facebook ads and privacy concerns—are in fact meager, more meager even than inflated dollars. What's worse, these connections are liabilities that we pretend are assets. We live on the Web in these hideous conditions of overcrowding only because—it suddenly seems so obvious—we can't afford privacy. And then, lest we confront our horror, we call this cramped ghetto our happy home!

Twitter is ten years old. Early enthusiasts who used it for barhopping bulletins have cooled on it. Corporations, institutions, and public-relations firms now tweet like terrified maniacs. The "ambient awareness" that Clive Thompson recognized in his early writings on social media is still intact. But the emotional force of all this contact may have changed in the context of the economic collapse of 2008.

Where once it was engaging to read about a friend's fever

or a cousin's job complaints, today the same kind of posts, and from broader and broader audiences, can seem threatening. Encroaching. Suffocating. Our communications, telegraphically phrased so as to take up only our allotted space, are all too close to one another. There's no place to get a breath in the Twitter interface; all our thoughts live in stacked capsules, crunched up to stay small, as in some dystopic hive of the future. Or maybe not the future. Maybe now. Twitter could *already be* a jam-packed, polluted city where the ambient awareness we all have of one another's bodies might seem picturesque to sociologists but has become stifling to those in the middle of it.

In my bolshevik-for-the-Internet days I used to think that writers on the Web who feared Twitter were just being old-fashioned and precious. Now while I brood on the maxim "Connectivity is poverty," I can't help wondering if I've turned into a banged-up street kid, stuck in a cruel and crowded neighborhood, trying to convince myself that regular beatings give me character. Maybe the truth is that I wish I could get out of this place and live as I imagine some nondigital or predigital writers do: among family and friends, in big, beautiful houses, with precious, irreplaceable objects.

The something lost in the design of the Web may be dignity—maybe *my* dignity. Michael Pollan wrote that we should refuse to eat anything our grandmothers wouldn't recognize as food. In the years I spent at Yahoo! News—not content-farming, exactly, but designing something on a continuum with click bait, allowing ads into my bio, and being trained (as a talking head) to deliver corporate propaganda rather than report the news—I realized I was doing something my grandmothers wouldn't have recognized as journalism. Privately I was glad neither of them had lived long enough to witness my tour of duty in that corner of the Web, doing Go-Gurt journalism.

RESPITE

Which brings me back to Hundreds and the other achingly beautiful apps, many of which could pass for objects of Italian design or French cinema. Shifting mental seas define the experience of these apps, as they do any effective graphic scheme in digital life, in which the best UX doesn't dictate mental space; it maps it. These apps caress the subconscious. The graphic gameplay on Hundreds seems to take place in amniotic fluid. The palette is neonatal: black, white, and red. The path through is intuition.

And this is strictly graphics. No language. Text here is deep-sixed as the clutter that graphic designers always suspected it was. The new games and devices never offer anything so pedestrian as verbal instructions in numbered chunks of prose. "If they touch when red then you are dead," flatly states a surreal sign encountered partway through Hundreds's earliest levels. That's really the only guideline you get on how Hundreds is played.

Playing Hundreds is a wordless experience. Even that red/dead line of poetry is more music than meaning. There's an eternity to the graphic swirl there; it's the alpha and the omega. "Death" would be too human and narrative an event to happen to the fog-toned circle-protagonists. These circles mostly start at zero. You drive up the value of the circles by touching them and holding them down, aiming each time to make the collective value of the circles total 100 before they run into an obstacle, like a circle saw.

Nothing about losing in Hundreds feels like dying. The music continues; the round can be replayed. No pigs (as in Angry Birds) or shirtless terrorists (as in Call of Duty) snort and gloat. You start again. Who says losing is not winning, and the other way around? In Hundreds even gravity is inconstant.

FRISBEE FOREVER

Digital, kaleidoscopic design can serve to undermine language. To deconstruct it. *Deconstruct* is still a frightening word, bringing to mind auteur architects and Frenchmen in capes. Here I use it to mean that digital design, especially in games, can call attention to the metaphors in language and teasingly demonstrate how those metaphors are at odds with language's straight-up, logical claims. So life and death are binary opposites? Not on Hundreds, which teaches the sublingual brain that life and death are continuous, world without end. Mixing up life and death in this way is, in fact, the operative principle of video games, as Tom Bissell's masterful *Extra Lives: Why Video Games Matter* convincingly argues.

Before the Internet, but presciently, Marshall McLuhan credited the world's new wiredness with dissolving binaries in the way of Buddhism: "Electric circuitry" (which elsewhere he calls "an extension of the human nervous system") "is Orientalizing the West. The contained, the distinct, the separate—our Western legacy—is being replaced by the flowing, the unified, the fused."

Where some game design breaks down language and the distinctions that undergird it, other design is tightly structuralist, instantiating boundaries and reminding players that they're contained, distinct, and separate. Frisbee Forever, a kid's game I'm choosing almost at random, works this way. A free candy-colored mobile game in which the player steers a Frisbee through a variety of graphic environments that look variously beachy, snowy, and Old-Westy, Frisbee Forever is one of those garish games at which some parents look askance. But the very week I downloaded Frisbee Forever for my then-six-year-old son, Ben, the Supreme Court ruled that video games were entitled to First Amendment protection, just like books, plays, and movies. I decided the game formally had redeeming value when I read Justice Scalia's words: "Video games communicate ideas—and even social messages—through many familiar literary devices (such as characters, dialogue, plot, and music) and through features distinctive to the medium (such as the player's interaction with the virtual world)."

So what's the idea—and even the social message—behind Frisbee Forever? The message is deep in the design: Never give up. Like many successful games, Frisbee Forever is built within a pixel of its life to discourage players from quitting—because if you quit, you can't get hooked. The game's graphic mechanics gently but expertly escort players between the shoals of boredom ("Too easy!") and frustration ("Too hard!"). This Scylla-and-Charybdis logic is thematized in the design of many popular app games (Subway Surfer, the gorgeous Alto's Adventure, and many of the so-called endless runner games). At PBS's website, for which educational games are always being designed, this protean experience is called "self-leveling." Tailored tests and self-leveling games minimize boredom and frustration so that—in theory, anyway—more people see them through.

This is certainly the logic behind Frisbee Forever. Just as a player steers her disc to keep it in the air, so Frisbee Forever steers her mood to keep her in the game. It's like a model parent. If a kid's attention wanders and his play becomes lackluster, the game throws him a curve to wake him up. If he keeps crashing and craves some encouragement, the game throws him a bone. Curve, bone, bone, curve. Like life.

And that's a potential problem. What's lost is bracing disorder, the spontaneous adaptations that lead to art and adventure and education. Frisbee Forever—and anything else self-leveling—conjures a fantasy world that's extremely useful when life's disorderly. But when things settle down in reality, the Frisbee game is *too* exciting. It does nothing to teach the all-important patience and tolerance for boredom that are central to learning: how to stand in line, how to wait at Baggage Claim, how to concentrate on a draggy passage of text. In fact self-leveling games suggest you never have to be bored. At the same time, Frisbee Forever is not nearly challenging enough. In real life you have to learn to tolerate frustration: how not to storm away when the pitcher is throwing strikes, how to settle for an Italian ice when sundaes are forbidden, how to try the sixth subtraction problem when you've gotten the first five wrong.

I find pleasing magic in the design of many digital and digitized games: Angry Birds, WordBrain, Bejeweled, Candy Crush. But I use their graphic worlds to keep myself safe from unstructured experience. To shut out mayhem and calm my mind. Often I find I want to keep the parameters of boredom and frustration narrow. I feel I need to confront rigged cartoonish challenges that, as it happens, you can—with pleasurable effort—perfectly meet. Games, like nothing else, give me a break from the feeling that I'm either too dumb or too smart for this world.

I'm not the only one in my demo. Thanks to the explosion of mobile games that have drawn in the crossword and Sudoku crowd, adult women now make up a bigger proportion of gamers (37 percent) than do boys eighteen and younger (15 percent), according to a study by the Entertainment Software Association. The *average* age of gamers is now thirty-five.

But of course I wonder what real challenges and stretches of fertile boredom, undesigned landscapes, and surprises I'm denying myself. And maybe denying my children.

SPRAWL

The schism between the almost fascist elegance of the sexiest apps, like Hundreds, and the chaotic-ghetto graphic scheme of the Web may have been inevitable. In the quarter-century since Tim Berners-Lee created the immensely popular system of hyperlinks known as the World Wide Web, the Web has become a teeming, sprawling commercial metropolis, its marquee sites so crammed with links, graphics, ads, and tarty bids for attention that they're frightening to behold. As a design object, it's a wreck.

There are two reasons for this. Two laws, even. And complain as we might, these two laws will keep the Web from ever looking like a Ferrari, *Vogue*, or the Tate Gallery. It will never even look like a Macintosh or an iPad, which is why Apple has taken such pains since the App Store opened to distance itself from the open Web, that populist place that is in every way open-source and to which we all regularly contribute, even if just with a Facebook like or an Etsy review.

1. The Web is commercial space.

The major links and sites are, of course, now paid for by advertisers, who covet click-throughs—or, better yet, taps of the "buy" button, which started to figure prominently on sites like Pinterest in 2015—and never stop fishing for attention. You think you're reading when you're on the Web; in fact you're *being read.* This is why the Web is now palpable as the massively multiplayer online role-playing game it's always been. You are playing the house when you play the Web, and the house is better at reading you than you are at reading it. To return to the bolshevik framework, *Read or be read* is today's answer to Lenin's old who-whom, *Who will dominate whom?*

The fact of this jockeying came home to me forcefully the first time Google introduced Panda, a series of changes to the company's search algorithm that reconfigured the felt experience of the Web. That's right: Panda influenced the whole Web. As surely as the graphic scheme of my desktop and gadgets is determined by Apple, the graphic scheme of my life on the Internet is determined by Google.

Before Panda was rolled out in 2011, the Web had started to look Hobbesian, bleak and studded with content farms, which used headlines, keywords, and other tricks to lure Web users into looking at video ads. Even after its censure by Google that dystopian version of the Web—as ungovernable content—is always in the offing. It's like demented and crime-ridden New York City: even after Giuliani and Bloomberg, we know that city could always come back.

Here's a flashback to the bonkers Web of 2011, as surreal as it sounds: Bosses were driving writers to make more words and pull photos faster and for less pay so they could be grafted onto video that came with obnoxious preroll advertisements. Readers paid for exposure to this cheaply made "media" in the precious currency of their attention. Prominent sites like Associated Content,

Answerbag, Demand Media, parts of CNN, part of AOL, and About.com (which was then owned by the *New York Times*) looked creepy and hollow, a zombie version of in-flight magazines.

"Another passenger of the vehicle has also been announced to be dead," declared one muddled sentence on Associated Content. "Like many fans of the popular 'Jackass' franchise, Dunn's life and pranks meant a great amount to me." This nonsense was churned out in a freelance, white-collar version of the Triangle Shirtwaist Factory. Many content-farm writers had deadlines as frequently as every twenty-five minutes. Others were expected to turn around reported pieces, containing interviews with several experts, in an hour. Some composed, edited, formatted, and published ten articles in a single shift—often a night shift. Oliver Miller, a journalist with an MFA in fiction from Sarah Lawrence, told me that AOL paid him about $28,000 for writing 300,000 words about television, all based on fragments of shows he'd never seen, filed in half-hour intervals, on a graveyard shift that ran from 11 p.m. to 7 or 8 in the morning.

Miller's job was to cram together words that someone's research had suggested might be in demand on Google, position these strings as titles and headlines, filigree them with other words, and style the whole confection to look vaguely like an article. Readers coming to AOL expecting information might soon flee this wasteland, but ideally they'd first watch a video clip with ads on it. Their visits would also register as page views, which AOL could then sell to advertisers.

A leaked memo from 2011 called "The AOL Way" detailed

the philosophy behind this. Journalists were expected to "identify high-demand topics" and review the "hi-vol, lo-cost" content— those are articles and art, folks—for such important literary virtues as Google rank and social media traction. In 2014 *Time* magazine similarly admitted to ranking its journalists on a scale of advertiser friendliness: how compatible their work was with advertising and the goals of the business side of the enterprise.

Before Google essentially shut down the content farms by introducing Panda to reward "high-value" content (defined in part by sites that had links to and from credible sources), the *Economist* admiringly described Associated Content and Demand Media as cleverly cynical operations that "aim to produce content at a price so low that even meager advertising revenue can support [it]."

So that's the way the trap was designed, and that's the logic of the Web content economy. You pay little or nothing to writers and designers and make readers pay a lot, in the form of their eyeballs. But readers get zero back: no useful content. That's the logic of the content farm: an eyeball for nothing. "Do you guys even CARE what I write? Does it make any difference if it's good or bad?" Miller asked his boss one night by instant message. He says the reply was brief: "Not really."

You can't mess with Google forever. In 2011 the corporation changed its search algorithm; it now sends untrustworthy, repetitive, and unsatisfying content to the back of the class. No more A's for cheaters. But the logic of content farms—vast quantities of low-quality images attached to high-demand search bait—still holds, and these days media companies like feverish BuzzFeed and lumbering HuffPo are finding ready workarounds for Google, luring people through social media instead of search, creating click bait rather than search bait, passing off ads as editorials. These are just new traps.

That's why the graphic artifacts of the Web civilization don't act like art. They act like games. I'm talking about everything from the navigational arrows to the contrasting-color links, the boxes to type in, and the clickable buttons. Rather than leave you to kick back and surf in peace, like a museum-goer or a flaneur or a reader, the Web interface is baited at every turn to get you to bite. To touch the keyboard. To click. To give yourself up: *Papieren!* To stay on some sites and leave others. If Web design makes you nervous, it's doing what it's supposed to do. The graphics manipulate you, like a souk full of hustlers, into taking many small, anxious actions: answering questions, paging through slide shows, punching in your email address.

That's the first reason the Web is a graphic mess: it's designed to weaken, confound, and pickpocket you.

2. The Web is collaborative space.

The second reason the Web looks chaotic is that there's no rhyme or reason to its graphic foundations.

In short order, starting in the 1990s, the Web had to gin up a universal language for design grammar. The result was exuberance, ad hoccery, and arrogance. Why? In 1973 the Xerox

Alto introduced the white bitmap display, which Apple promptly copied with the Macintosh's bellwether graphical interfaces. As a verbal person, I feel some nostalgia for what might have been. The evocative phosphor-green letters on a deep-space background that I grew up with gave way to the smiley Mac face and the white bitmap that turned computing entirely opaque. After I saw the Mac I lost interest in learning to code. I was like an aspiring activist who, before even getting started, was defeated by a thick layer of propaganda that made the system seem impenetrable.

In literary critical terms inherited from the great Erich Auerbach in *Mimesis*, the grammar of the computer interface would go from parataxis—weak connectives, like all that black space, which allowed the imagination to liberally supply and tease out meaning—to hypotaxis, in which hierarchies of meaning and interpretive connections are tightly made *for* a user, the visual field is entirely programmed, and, at worst, the imagination is shut out.

When I switched to a Mac from my Zenith Z-19 dumb terminal, called "dumb" because it had nothing in its head till I dialed in to a mainframe, I bore witness to the dramatic transition from phosphor to bitmaps. Gone was the existential Old Testament or *Star Trek* nothingness of those phosphor screens—you can picture them from *War Games*—which left you to wonder who or what was *out there*. The new, tight, white interface snubbed that kind of inquiry and seemed to lock you out of the "friendly" graphical façade. It was as though a deep, wise, grooved, seductive, complex college friend had suddenly been given a face-lift, a makeover, and a course in salesmanship. She seemed friendly and cute, all right, but generically and then horrifyingly so. Nothing I did could ever bait her into a free-flowing, speculative, romantic, melancholic, or poetic relationship ever again.

As for coders, they have known since the bitmap appeared that rectangular screens, indexed by two coordinates, would demand design. And they were thrilled. The first design could be written this way: N46 = black; F79 = white; and so on. Though vastly more expensive, the bitmap display was greedily embraced by the computer companies of the 1960s and 1970s for a significant reason: coders hugely preferred its grid and iconography over linear letters and numbers.

But why? For the answer I had to look to the testimony of coders on the subject, and I found that the profession's preference for graphics and iconography over straight text has to do with cognitive wiring. Like Nicholas Negroponte, the dyslexic founder of MIT's Media Lab who as a child preferred train schedules to books, and Steve Jobs, who liked calligraphy more than words, many computer types shun narrative, sentences, and ordinary left-to-right reading. Dyslexic programmers, not shy about their diagnosis, convene on Reddit threads and support sites, where they share fine-grained cognitive experiences. A sample comes from a blog post called "The Dyslexic Programmer" by a coder named Beth Andres-Beck:

My dyslexia means that the most important thing for me about a language is the tool support, which often rules out new, hip languages. It took me a while to figure out that my dyslexia was the reason I and the command-line centric programmers would never agree. I've faced prejudice against non-text-editor programmers, but often only until the first time they watch me debug something in my head. We all have our strengths ;-)

"Debugging," according to the educational theorist (and notable dyslexic) Cathy N. Davidson in *Now You See It*, is actually an *agricultural* skill that may even be at odds with traditional literacy. It allows farmers to look at a field of alfalfa and see that three stalks are growing wrong and the fertilization scheme must be adjusted. This is a frame of mind known to coders like Andres-Beck, who can identify the bug in vast fields of code without seeming to "read" each line.

For debuggers, as Davidson makes plain in her book, traditional text is an encumbrance to learning. And as Andres-Beck observes, command-line interfaces that use successive lines of text, like books, also bedevil those who find reading challenging. More symbolic, spatially oriented "languages" begin to seem not just friendlier to this cognitive orientation; they look progressive. In 1973, when programmers glimpsed the possibilities of the Xerox bitmap, they never looked back.

This kind of interface defied the disorientation that had long been induced by letters on paper. Torah scribes in the first century defined literacy as the capacity to orient oneself in tight lines of text on a scroll without whitespace, pages, punctuation, or even *vowels* to mark spots for breath or other spatial signposts. A real reader in ancient times—and there weren't many—was expected to mentally go a long way to meet a text. To many programmers—with a visual, agri-debugger's intelligence at sharp odds with this practice—this kind of literacy seems ludicrous, even backward. ("Once you use a structurally aware editor going back to shuffling lines around is medieval," writes one satisfied customer of nontext coding on Reddit.)

As Negroponte explains in *Being Digital*, many digital natives, and boomers and Gen Xers who went digital, are drawn to the jumpy, nonlinear connections that computer code makes. On the site io9 recent studies of the differences between Chinese dyslexics and English dyslexics were used to make the case that dyslexics make good programmers, as programming languages contain the symbolic, pictographic languages that many dyslexics prefer.

In the 1970s and 1980s bitmaps were welcomed like a miracle by PC programmers. Writers sometimes long for the low-graphic blackout screen in the beloved and lo-fi word-processing app WriteRoom, but no one else seems to. The majority of computer users, and certainly programmers, were overjoyed at the reprieve from traditional literacy that the bitmap granted them.

It's important to realize too that early programmers were especially ready to give up writing *manuals*, those byzantine booklets that Xerox, IBM, and Compaq workers had to come up with when it came time to explain their esoteric hobby to the noncoding PC crowd. The 1980s notion of user-friendliness was really a move from that tortured lost-in-translation technical language to *graphics*, which in theory would be legible in all languages and to a range of cognitive styles. Good graphics could ideally even move us away from the parody-worthy manuals to self-explanatory ("intuitive") interfaces that would require no instruction at all.

The English designer Jon Hicks of Hicksdesign, which has created iconography for Spotify and emoticons for Skype, spells out some of the Web's absolute dependence on pictorial language. He even uses the word *miracle*: "Icons are more than just pretty decorative graphics for sites and applications, they are little miracle workers. They summarize and explain actions, provide direction, offer feedback and even break through language barriers."

The original graphics on bitmaps were very simple—but not elegant. Initially the screen could show only a binary image, which worked like those children's games where you pencil in one square of graph paper at a time in order to produce a black-and-white picture. Before that, hinting at what was to come, all kinds of fanciful typography covered my Zenith Z-19 terminal screen in the 1970s and early 1980s, when sysprogs would draw faces in dollar signs or "scroll" people, sending them line after line of random symbols to annoy them and clog their screen. That trolling use of graphics—to confront and annoy—is still in effect in various corners of the Web.

Today a bitmap is really a pixmap, where each pixel stores multiple bits and thus can be shaded in two or more hues. A natural use of so much color is realist forms like photography and film. But the profusion of realist imagery since Web 2.0, when wider bandwidth allowed photos and videos to circulate, can lead us to overlook the graphics that constitute the visual framework and backdrop of almost any digital experience.

The exuberance with which programmers, stymied by straight text, embraced graphics is the second reason the Web is a graphic wreck. It was made by manic amateurs trying to talk in pictures, not by cool pros with degrees in Scandinavian design.

THE REFRACTED RECORD OF TECH HISTORY

Right now, looking at my laptop screen, I see a row of tight, fussy little icons below the Google doc I have open. This is the "deck" on a Mac. I used to know these like my own five fingers, but that's when I only *had* five icons. Now there are—twenty? These represent the tips of the icebergs for the major tech players, and like ticker symbols, they all jostle uncomfortably for my attention. I used to think of Microsoft Word as the blue one, but now I see that Apple has seized various shades of what the humorist Delia Ephron disparages as "bank blue," and the iMessage app (with cartoon talk bubbles) stands out in that shade, along with the Mail stamp-shaped icon, compass-shaped Safari, and protractor App Store. None of these has a word on them. They look maddeningly alike.

The Microsoft Office icons at least look like letters, each made of a single shaded satin ribbon: W, P, X, O. That's Word, PowerPoint, Excel, Outlook. The spindly shape and contemporary colors—bank blue, yes, but also burnt orange, grass green, and yellow—are a legacy of the ferocious determination of Microsoft to set itself apart from Apple and to seem less isolated and self-contained (like circles) and more compatible and connected (like cursive letters). These Microsoft apps, which I paid dearly for since, because of the Shakespearean Jobs-Gates antagonism, they don't come standard with Macs anymore, *would* set Office apart from my machine's native apps were it not for the fact that Google introduced a virtually identical palette some years ago in its multihued Chrome and Drive icons.

So there's the motley lineup. And I haven't said a word about FitBit, Skype, Slack, Spotify, and whatever else is down there. This is hardly a team of miracle workers. It's more like a bagful of foreign coins. And though no doubt the great icon designers of our time created them, they're now no more emotionally striking or immediately legible than any other timeworn pictographic alphabet.

BLADE RUNNER

Click on Chrome, though, and you're zapped onto the Web. Only then is it clear that the Mac interface, even with its confounding icons, is, in contrast with the Web, a model of sleek organization.

The Web is haphazardly planned. Its public spaces are mobbed, and urban decay abounds in broken links, ghost town sites, and abandoned projects. Malware and spam have turned living conditions in many quarters unsafe and unsanitary. Bullies, hucksters, and trolls roam the streets. An entrenched population of rowdy, polyglot rabble dominates major sites.

People who have always found the Web ugly have nonetheless been forced to live there. It is still the place to go for jobs, resources, services, social life, the future. In the past eight or ten years, however, mobile devices have offered a way out, an orderly suburb that lets inhabitants sample the Web's opportunities without having to mix with the riffraff. This suburb is defined by those apps: neat, cute homes far from the Web city center, many in pristine Applecrest Estates—the App Store.

In the migration of dissenters from those ad-driven Web sites, still humming along at www URLs, to pricey and secluded apps we witnessed urban decentralization, suburbanization, and the online equivalent of white flight: smartphone flight. The parallels between what happened to Chicago, Detroit, and New York in the twentieth century and what happened to the Internet since the introduction of the App Store are striking. Like the great modern American cities, the Web was founded on equal parts opportunism and idealism.

Over the years nerds, students, creeps, outlaws, rebels, moms, fans, church mice, good-time Charlies, middle managers, senior citizens, starlets, presidents, and corporate predators flocked to the Web and made their home on it. In spite of a growing consensus about the dangers of Web vertigo and the importance of curation, there were surprisingly few "walled gardens" online like the one Facebook once purported to represent. But a kind of virtual redlining took over. The Webtropolis became stratified. Even if, like most people, you still surf the Web on a desktop or laptop, you will have noticed paywalls, invitation-only clubs, subscription programs, privacy settings, and other ways of creating tiers of access. All these things make spaces feel "safe" not only from viruses, instability, unwanted light and sound, unrequested porn, sponsored links, and pop-up ads but also from crude design, wayward and unregistered commenters, and the eccentric voices and images that make the Web constantly surprising, challenging, and enlightening.

When a wall goes up, the space you have to pay to visit must, to justify the price, be nicer than the free ones. The catchphrase for software developers is "a better experience." Behind paywalls like the ones that surround the *New York Times* and the *Wall Street Journal* production values surge. Cool software greets the paying lady and gentleman; they get concierge service, perks. Best of all, the advertisers and carnival barkers leave you alone. There's no frantically racing to click the corner of a hideous Philips video advertisement that stands in the way of what you want to read. Those prerolls and goofy in-your-face ads that make you feel like a sitting duck vanish. Instead you get a maitre d' who calls you by name. Web stations with entrance fees are more like boutiques than bazaars.

Mobile devices represent a desire to skip out on the bazaar.

By choosing machines that come to life only when tricked out with apps, users of all the radical smartphones created since the iPhone increasingly commit themselves to a more remote and inevitably antagonistic relationship with the Web. "The App Store must rank among the most carefully policed software platforms in history," the technology writer Steven Johnson observed—and he might have been speaking conservatively. Policed why? To maintain the App Store's separateness from the open Web, of course, and to drive up the perceived value of the store's offerings. Perception, after all, is everything; many apps are to the Web what bottled water is to tap: an inventive and proprietary new way of decanting, packaging, and pricing something that could be had for free.

Jobs often spoke of the corporate logos created by his hero Paul Rand (creator of logos for IBM, UPS, and Jobs's NeXT) as "jewels," something not merely symbolic but of pure value in themselves. Apps indeed sparkle like sapphires and emeralds for people enervated by the ugliness of monster sites like Craigslist, eBay, and Yahoo!. That sparkle is worth money. Even to the most committed populist or Web native there's something rejuvenating about being away from an address bar and ads and links and prompts, those constant reminders that the Web is an overcrowded and often maddening metropolis and that you're not special there. Confidence that you're not going to get hustled, mobbed, or mugged—that's precious too.

ANGER

But not all app design can or should be chill. Some of it should be provocative, emotional, even enraged. I like to think of Rovio, the game studio that created the juggernaut Angry Birds, as the center of rage-based gaming. The narrative of Rovio's masterpiece—that the pigs have stolen your eggs, your babies, and "have taken refuge on or within structures made of various materials"—is just superb. Reading Wikipedia's summary of Angry Birds just now makes me want to play it again. Though of course the cravings should have subsided years ago, when I was on the global top-1,000 leaderboard for Angry Birds. (Autographs available on request.)

What a great pretext for a game: pigs steal your babies and then lodge themselves in strongholds made of stone, glass, or wood. They take refuge, as in "the last refuge of a scoundrel," and the refuges are maddeningly difficult to penetrate or topple.

In Angry Birds, as so often in life, the material world seemed to have conspired to favor the jerks, endowing them with what looked like breastworks, berms, and parapets, as if they were the beneficiaries of some diabolical foreign-aid package. Those gross, smug, green pigs stole my flock's *babies*, and they're sitting pretty in stone fortifications that they didn't even build themselves. And the looks on their fat faces? Perfectly, perfectly self-satisfied.

Think you're too good for me, eh? That you'll rob me and I'll just be *polite about it*? You have your elaborate forts and your snorting equipoise. I have nothing but my sense of injury. My rage. And so I take wobbly aim at them, the pig thieves, in Rovio's world without end, in which there are hundreds of levels to master and the game gets bigger and bigger with constant updates.

Angry Birds is a so-called physics game, which suggests education, and also a puzzle in reverse, as you must destroy something by figuring out how its pieces come apart. Your tools are these birds: the victims of the theft but also your cannon fodder. Each bird that is launched dies. Though there is no blood, as it is death by cartoon *poof*, every mission is a suicide mission.

ELEGANCE

At the other end of the design continuum is Device 6, a hyper-polished game in the spirit of Hundreds. In this one, though, the player is a reader at heart. That's too bad because the arts on the Internet are still brutally segregated, and no one has yet been able to bring together beautiful prose and beautiful design.

The game, then, is an interactive book, with prose that branches off like a cross between e. e. cummings and "Choose Your Own Adventure." The protagonist is Anna, a gutsy woman with a hangover and a headache. The player tracks her as she wakes up from a blackout in a shadowy castle. She prowls around trying to figure out what's up and how to get the heck out by answering riddles, cracking codes, and solving math problems that are in fact arithmetic or the simplest algebra.

The game's ambience is ingenious, evocative, and chill without being precious and "overdecorated" (as Anna observes of the castle). It's no surprise that this elegance comes from design-delirious Sweden, where in 2010 Simon Flesser and Magnus "Gordon" Gardebäck formed Simogo. The game has something of Ingmar Bergman's beloved "world of low arches, thick walls, the smell of eternity, the colored sunlight quivering above the strangest vegetation of medieval paintings and carved figures on ceilings and walls."

The sprawling castle is a hoardery pile jammed with broken, weird, outmoded tech, including Soviet-era computers. The graphics evoke the 1960s, the cold war, and James Bond, but the photographic element of the game also expresses 1960s nostalgia for the 1930s and 1940s. This is a neat trick that only confident European design fiends could pull off. At one point Anna comes upon an R&D lab that makes toys or weapons or weaponized toys. It's creepy without being too on-the-nose horror flick. The chief drawback to the game's being made by Swedes is that the English-language writing in Device 6 is plodding. Like the language in too many games and apps, the prose here is a placeholder—not exactly Agatha Christie or Alan Furst.

With some experimental exceptions, like Hundreds, graphic design cannot exist in a vacuum online. On the Web it continues to jostle uneasily and sometimes productively with sound, text, photography, and film. Naturally designers prefer to the open Web the hermetic spaces of apps, which they have much more control over and which aren't built on often ugly hypertext. But these designers can't seal themselves off forever from the challenge of shared space, referring in their work to offline print design and creating mute, pantomime realms of immaculate shapes. Digital design will find itself when the designers embrace the collaborative art of the Internet and join forces with writers, sound designers, and image-makers.

2

TEXT

The history of digitization is the history of reading. Ada Lovelace's use of Bernoulli numbers for computers evolved steadily into natural-language code like BASIC, and then into plain words, like *hello*. Before long came the kindergarten staples of the mobile, social Web: *apple*, *share*, and *friend*.

Code that moves only forward is called "write." Code that can be reviewed and remembered, or can review and remember itself, is called "read." Unlike real-life experience, which is write-only and cannot (yet) be paused or rewound, all of the Internet—except in its most secure, theoretical, and self-destroying corners, and mostly even there—can be read. You can review, search, index, and store the best and worst of what's been digitally done and said in the world.

In the days of newsprint and broadcast we used to fear the disposability and ephemerality of our communications. Now we fear their indelibility and susceptibility to interception. Above all else, the Internet is *readable*.

Another principle of digital civilization is *No human can read it*. Digital literacy therefore involves chiefly the refusal to read. (Schopenhauer: "The art of *not* reading is a very important one. It consists in not taking an interest in whatever may be engaging the attention of the general public at any particular time.") Shrewd digital readers build defenses against reading and strenuously resist hyperlexia as the plague of our time. It's hyperlexia that keeps people's eyes fixed on their phones and not on nature, art, friends, mates, children, or work. And it's hyperlexia that leads to fatalities in driving-while-texting accidents. We have become so compulsively unwilling to stop reading (Facebook, Tinder, WhatsApp) that we will risk our lives, livelihoods, and certainly marriages to keep at it. At the same time, curiously, the keepers of the canon of media—books, magazines, newspapers—are hesitant to legitimize digital reading and often blind themselves to it. To that crowd, what hundreds of millions of us do with near-murderous ardor frequently doesn't even register as reading.

The very first concern I heard expressed about the World Wide Web, in 1991, touched on this phenomenon. It came from John Perry Barlow, the Grateful Dead lyricist, rancher, and organizer who is otherwise a boisterous partisan of digital life, having been a founding member of the hallowed proto-feminist Internet community known as the Well and the founder of the Electronic Frontier Foundation, which brought Berkeley-style fundamentalism about free speech to digital space. (I was in awe at first sight; later I sat with Barlow in Austin, Texas, where he told me about his first use of digital technology, to organize Deadhead meetups.)

Reading the Internet, Barlow said, was like "drinking from a fire hose." The analogy wasn't his, or maybe it was; in either case the image was vivid. Sitting awkwardly in the brand-new Harvard Cyberlaw Seminar, which later became the Berkman Center for Internet and Society, I instantly imagined how I'd quench my thirst if confronted with nothing but a water cannon. (Bursting at some 80 bars of pressure, the fire hose, like the first analog computer, was invented in ancient Greece.) I remember resolving that I'd seek the outermost periphery of the water blast and aim to catch some spray on my tongue, and that I'd not let anxiety over dehydration lead me to risk a water-blast beheading. One thing I was not going to do was run.

When the Internet, with its world-historical collective cerebral cortex, so thoroughly seems to overpower the human brain, it's no wonder that our own mortal brains have come to seem—and maybe be—terminally deficient and damaged. We're reading as fast as we can and reading far more than is needed to gain life instructions or quench intellectual thirst, and yet rather than regularly gasp in awe at the Alpine—or, no, extraterrestrial sublimity of the Internet, we seem intent on taking the measure of our own stuntedness, which we imagine as newly revealed.

The Internet has taught us that we're not studious, focused, or learned enough. Or is it that we don't have enough time in nature, or that we bowl alone? In increasingly attenuated ways we lay blame for our seemingly new limitations as "readers"—another name for moral agents in a world of symbols—at the feet of the Internet, in the mirror of which we first constructed those limitations.

Google's stated mission "to organize the world's information" echoes the founding tale of Western ambition: Faust's deranged desire for unlimited knowledge. To keep pace with the Internet, we'd have to forfeit our souls. But that would only be step one. By step two we'd be licked, with no chance of knowing what Google knows. Licked, that is, and now soulless.

We have, in short, come up with an impossible task for ourselves and now reflexively address our inadequacy with the brain cell–bruising pastime known as "worrying." Not long ago the Edge Institute asked me to contribute an answer to its yearly question for scientists and philosophers. "What do you think about machines that think?" was the confidential query. What indeed! My answer came easily:

> Outsourcing to machines the many idiosyncrasies of mortals—making interesting mistakes, brooding on the verities, propitiating the gods by whittling and arranging

flowers—skews tragic. But letting machines do the *think-ing* for us? This sounds like heaven. Thinking is optional. Thinking is suffering. It is almost always a way of being care-ful, of taking hypervigilant heed, of resenting the past and fearing the future in the form of maddeningly redundant internal language. If machines can relieve us of this oner-ous non-responsibility, which is in pointless overdrive in too many of us, I'm for it. Let the machines perseverate on te-dious and value-laden questions about whether private or public school is "right" for my children; whether interven-tion in Syria is "appropriate"; whether germs or solitude are "worse" for a body. This will free us newly footloose humans to play, rest, write, and whittle—the engrossing flow states out of which come the actions that actually enrich, enliven, and heal the world.

POETRY AND TWITTER

Does poetry matter? asked the *New York Times*. Or has Twitter killed it? What a question. Asking what's to become of poetry in the age of Twitter is like asking what's to become of music in the age of guitars. It thrives. It more than thrives; it grows metastati-cally, invasively, inoperably. Poetry on the Internet has shot far past relevancy through indispensability and finally to vaporiza-tion. Poetry is the air we damn well breathe.

The curiously formal verse of Twitter is in general neither good nor bad. That doesn't hurt its status as poetry: it is language precisely and even artfully deployed. This poetry loses and gains jobs, esteem, and reputation. Wars, rumors of war, the fates of men and women hang in its lyrical balance. It costs, in short, and it pays. This is what relevancy is, maybe harder to define than poetry. Tweets are news. They are history.

@melissabroder once tweeted, "Review of leaving the womb: ★☆☆☆☆." "I told my wife the water is too shallow. She said, wait 'til you get to know it better," tweeted the language poet @chrlesbernstein not long ago. @crunk_bear was fired after she tweeted, "Naked. Wet. Stoned," which might have been a line in a minor Allen Ginsberg poem. I'd end up shilling for a larky social service that entirely by accident came to double as a massive publisher, ballast and megaphone for the world's aspiring poets—which is to say, it turns out, everyone.

Will the "world"—Bostonians, Manhattanites—ever again wait for the publication of the next Carl Sandburg poem in the *Saturday Evening Post*? Will fans throng to hear Ted Berrigan read his latest from *Fuck You: A Magazine of the Arts*? Unlikely, though the great Paul Muldoon says an Isle of Wight–style crowd jammed Fire Island not long ago to hear him and others read original and Frank O'Hara poems. But how in the world could the waning of the Iowa-*Ploughshares* regime mean that poetry itself—the musical and idiosyncratic use of language—is washed up? And yet the case is made, again and again, that poetry is obsolete because those beloved analog platforms are in decline.

If Twitter is allowed to be poetry, it's as though a special dispensation has been granted. Curiously this is not because tweeters are coarse, ill-intentioned, or untrained. It's because tweets are short. Though many of us spend more time on single tweets than on hasty email—composing, compressing, tagging, revising as the context evolves—the fact that tweets are quickly read is somehow evidence they're as trivial as birdies, even eerily brain-damaging. (As night follows day the shortness of tweets conjures the standing sophistry around our "attention span," an occult feature of the mind that in theory is as fixed as a boxer's reach. But the attention span is always invoked as a thing deformed and on closer examination surely does not exist, except as always already julienned.)

Lyric poetry has always been short. That's why it's not, for example, epic. To plenty of poets in plenty of languages, 140 symbols is expansive. Confucius's adages were rarely longer than twenty Chinese characters; 140 would have seemed longwinded. Ace Confucianisms are brief even transliterated: "What you do not want done to yourself, do not do to others." That first draft of the Golden Rule, composed in 460 BC or so, sings in any character set.

Blaise Pascal's *Pensées*, published in 1669, were also short. "Do you wish people to believe good of you? Don't speak" is fifty-eight characters, and about the same in French. "Too much and too little wine. Give him none, he cannot find truth; give him too much, the same." That's ninety-seven. In American English, Ralph Waldo Emerson is our leading epigrammatist. "Adopt the pace of nature: her secret is patience" consumes a scant forty-nine characters. "Have the courage not to adopt another's courage" is forty-eight.

Tweets are not diseased firings of glitchy minds. They're epigrams, aphorisms, maxims, dictums, taglines, headlines, captions, slogans, and adages. Some are art, some are commercial; these are forms with integrity. True, a single tweet is not useful for distracting people from plague and war for hours and days, like tales by Boccaccio or Scheherazade. (Though, as Kate Lee once pointed out to me, a perpetual Twitter feed tends to serve us well in times of lockdown emergencies.) A tweet or two does not fill enough pages to give a chunk of biblio-merchandise heft to justify a $20 price tag. Nor do they even fill the whitespace of a perfect half-page of a saddle-stitched magazine so ads can be sold on the rest of that page for a nice, round price.

But the blank refusal of tweets to answer to the exigencies of earlier eras: that would seem to make them *more* poetic, not less.

BOMBAST

Even as wordless videos, without dialogue, subtitles, or voice-over—music videos, natural anomalies, freak accidents, fistfights, dances, stunts, sleights of hand, animal antics—attract enormous polyglot audiences on YouTube, viewers seeking interaction and discussion turn to text-heavy polemical clips about religion or politics. These videos draw smaller but infinitely more voluble audiences than the dumb shows.

A video called *The Truth about Islam from an Ex-Muslim Lady*, which shows a woman on a TV news program delivering a fearsome disquisition in Arabic on Samuel P. Huntington's clash-of-civilizations idea, was for a long time the most discussed video on the site. The woman, identified as an Arab American psychologist, celebrates the "civilization" of the West and denigrates the "backwardness" of Islam, according to the English subtitles. Something about this video clip inspired viewers to lay bare their ideological ids, and in various forms it has prompted hundreds of thousands of comments.

Similarly *Atheist*, which intersperses anti-atheist Bible passages with images of illustrious nonbelievers, has occasioned a sprawling argument about superstition, eschatology, and scientism among viewers with screen names like tylenolalcohol and blindedbynoise. The historical pedantry on display in another much discussed clip, *Macedonia Is Greece*, in which placards of factoids and photos of regional maps are marshaled to contend that no part of the former Yugoslavia should call itself Macedonia, triggered a flood of responses in Serbo-Croatian and Greek. Ancient vendettas thrive in this commentary. A cruder video, *Kuran ve Tuvalet Kagidi* (*My FREEDOM of SPEECH*), apparently created and posted by a secular nationalist Turk now living in Ivory Coast, proposes that toilet paper is more useful than the Koran. The sixty thousand responses to the brutish video are almost exclusively in Turkish.

The shouting match on YouTube kicked off by *The Truth about Islam* is perhaps the most instructive about digital rhetoric: part atavistic race riot, part religious disputation, and part earnest effort at enlightenment, the expansive commentary is a full-blown novel of world religion that dramatizes the fascinating and often shocking preoccupations of today's desk-chair ideologues.

The clip itself has a complicated provenance. A title sequence indicates that the first version is a presentation of a presentation of a presentation—YouTube clips often come in double if not triple sets of quotation marks—and the presenters are not exactly politically neutral. The segment originally appeared on Al Jazeera, after which it was excerpted, subtitled in English, and posted to the Internet by the Middle East Media Research Institute (Memri), an organization founded by a former Israeli national security adviser. (While the American and European media rely on Memri's conscientious translations of documents from the Middle East, critics complain that the organization disseminates only alarmist material about the Arab world.) The video was uploaded to YouTube by someone whose blog is unsubtly titled BanIslam. That version of the video (with its long commentary) was evidently taken down and then reposted by someone angry that "western halal-hippies" had tried to censor it. Comments are disabled for the current iteration, so I refer to my exegesis on the first set.

After the video's title sequence, the "ex-Muslim lady," a dour-looking brunette named Wafa Sultan, begins an incantation. "The clash we are witnessing around the world is not a clash of religions," she says, according to the subtitles. "Or a clash of civilizations. It is a clash between two opposites, between two eras." Praising Jewish accomplishment and Buddhist pacifism and heaping scorn on Muslim barbarism, she speaks in a voice that literally resounds; the Al Jazeera set echoes like an amphitheater. She praises Western culture in the ferocious cadences usually associated with the mullahs who condemn it. "They started this clash," she says of Muslims, "and began this war!"

After a wave of approval from commenters when the video first appeared, skeptics soon arrived, and debate raged. Some commenters wondered aloud: Was Sultan a paid propagandist? An anti-Muslim poster named Crusader18 (no less) stirred the pot, flaunting a kitschy eighteenth-century prose style and challenging others to justify the Koran. As he put it, "I judge Islam by it's [*sic*] original Texts and the actions of the Salafists . . . Poison Fruit from a Poison Tree." He continued, "DEFEND ISLAM you Cutthroat! CAN YOU?" While some commenters responded by delving into Koranic exegesis, a viewer named MassLax pushed back from another direction, suggesting that Crusader18 and his ilk secretly envied Muslim solidarity: "One billion people from a vast range of races, nationalities and cultures across the globe—from the southern Philippines to Nigeria—are united by their common Islamic faith. The Kufars are jealous. God willing they will die in their jealousy."

This level of imperious, along with the anachronistic locutions, is the house style of political YouTube. (Those familiar with it from YouTube were not entirely shocked when it surfaced with a vengeance in the campaign rhetoric of Donald Trump.) The Internet buffs who savor the theatrics of formal debate seem to have grown up on science fiction and courtroom dramas, as well as the Bible or the Koran. Several boast of owning, enslaving, and burying their opponents with wit. They also call each other "fools," "zygotes," "sophists," "tumors," "ghouls," and "voles." In one wounding exchange Crusader18 says to budavol, a tenacious spokesman for secular values, "Your bigotry towards Christians leads me to believe you may have been molested by a priest. . . . I hope he didn't catch anything from you."

Harsh. Certainly this is not for all audiences. But as a commenter named AuraX says of the message board, "This is the battlefield," a rabbley showdown that positions itself along the fault lines of the world's great debates.

The commentary may come to seem enervating, hundreds of micro-arguments effectively under the school-yard heading "Whose Religion Is Better?" Certainly the anarchy of the Internet can cause disorientation: no final arbiter, no master author will settle even the most trivial matter of fact, let alone summarize the terms of an argument or furnish an illusion of finality. But isn't commentary on commentary on commentary always the way of religion? This pompous Augustan-style rhetoric in sinuous YouTube is only the latest way of conducting traditional disputations.

LANGUAGE INFLUX

The words just flow into your brain. Was it me, or did ecstasy seem to buzz through that sentence? The CEO of Spritz, Inc., Frank Waldman, was describing the company's new reading app, Spritz.

What a promise! Reading without so much as trivial ocular effort. The ultimate in sedentary pleasures. Reading exertion zero. No eye movement, no tracking lines, no keeping one's place on a page.

An application that strategically flashes one word at a time to help you pick up your reading pace, Spritz sounded hot to me—especially that greedy and passive language influx. "Flow" turned out to be an evocative image for others too. Reviewers couldn't wait to try the service, which made exorbitant claims to "reimagining reading" and had been three years in the making in Boston, the nation's undefeated capital of learning and literacy. But they were chiefly eager to debunk Spritz as an unctuous fraud. Spritz, wrote Ian Bogost in the *Atlantic*, is "the latest repackaging of a decades-old optical snake oil."

That wasn't all. In another insta-debunk that appeared before the app's official launch, the *New Yorker* quoted Keith Rayner, a psychology professor with an eye-tracking lab at UC San Diego, on the science of Spritz. "Hogwash" was his finding. Snake oil *and* hogwash: a neat trick. Calling bullshit on new tech in blustery terms never fails to make tech-blog headlines.

The Spritz prelaunch buildup and takedown marked a cartoon data point in the digital revolution. Unexploited as of this writing, with an entirely reasonable ambition, Spritz offers a new way to read quickly. There's nothing indecent in this offering, but the dialogue around Spritz was hardly like one that might attend a material-world invention: a new way to store bath toys or trap mice. Instead the reaction to Spritz was near-hysterical.

Like iTunes, YouTube, and Instagram, Spritz is a digital artifact that taps into both the revolution's otherworldly promises
and its profound sensory-emotional losses. Sure enough, the
Spritz inventors speak of their product as if it is revealed religion. Spritz, the materials claim, "reimagines" and "reinvents"
reading. Reading: no less than the defining act of history, the
humanities, and, some say, humanness itself.

The mourners' protests and sobs were just as exaggerated:
Spritz is snake oil and hogwash and a dire threat to all that is
sacred about symbols and sentience. Bogost's enormously evocative review proposes that Spritz might offer not a way to read
but "an attentional version of data compression," required because the Internet is too vast and fast to be legible. Rather than
reading we're encountering data and tallying those encounters
as cultural capital. "The faster we can be force fed material," as
Bogost cynically puts it, "the larger volume of such matter we
can attach to our user profiles and accounts as data to be stored,
sold, and bartered."

Spritz is for not-real readers, evidently. And here's where
our most sacred class values come in, pounded with a mallet. It's
no surprise that the *Atlantic* and the *New Yorker* serve as the old
guardians, policing the borders of literacy. Spritz works, they
concede, for stuff you have to read—discovery, briefs, memos,
and social media "updates" for data merchants and info trades-
men—but not for the pleasure reading of books that defines the
bona fide man of leisure and letters. Juxtaposing a moral line on
a class line, Spritz, several reviewers argue, is not for virtuous

people who like to read. It is for subliterate business types who have to read.

Separating real from false reading, and real from false readers, has been a power proposition with sinister consequences since the first century AD, when *sofers* argued that reading the then-new codices (books with separate pages) wasn't really reading. Until you've found your way in a maddeningly disorienting Torah scroll, went the argument that safeguarded the scribes' elevated status, you haven't really read at all. (I thank the novelist Dara Horn for this point.)

But then on Twitter, DeSales Harrison, a poet and professor of English at Oberlin, opened my eyes to a credible loss Spritz might represent. He tweeted his review: "Speed reading apps deny the existence of white space, line breaks, the idea of the page." He framed the pain this way: "#deathofpoetry."

What I understood suddenly was that Spritz was the latest way for technologists to flog the magic of being digital and the latest way for analog minds to register one of the most painful insults to our cognitive organization. Like schools that abolish recess, the Internet often seems to deny its citizens R&R— spaces (white space, line breaks) for response, intake, and contemplation. For recess. It's no wonder that a discourse around "mindfulness" and meditation has grown up in response to a digital world of wall-to-wall stimulus.

The old drinking-from-a-fire-hose charge acquired new momentum with Spritz. No time to swallow! We don't want words to course without cease into the brain, the poets said. We need to sit still in the wordlessness, our brains unmolested even by seemingly effortless inward flow. We want to actually drink, to taste the liquid on our tongue, not just to be hydrated by the machine, as by an IV. And we want, at regular intervals, to feel that our thirst is being quenched, *is* quenched.

That poetry on the page, with its crucial spatial organization, seemed under fire by Spritz made sense as a pretext for grief and longing. Poetry, spoken or written, is always dying, supplanted by prose, an early form of "false" literacy; poets exist in part to eulogize verse itself. But most of the attacks on Spritz came dressed in scientific jargon. That only made them more off-kilter. In spite of Spritz's research showing that readers using it retained as much or more of what they read using traditional technologies (pages, books, scrollable screens), for example, John M. Henderson, director of something called the Institute for Mind and Brain at the University of South Carolina, cited other research on similar (but not identical) tech from the 1970s. With an application like Spritz, Henderson inferred from his decades-old data, "there just isn't enough time to put the meaning together and store it in memory (what psychologists call 'consolidation')."

Consolidation. At times it surprises me that pre-neuroscience psychology still insists on its own technical-sounding names for various kinds of thinking. At the same time, these names—like Freudian terminology—rarely seem to designate anything in the visible world. The nonostensive jargon, rather than illuminating anything, adds another muddy layer of (typically worse) poetry to cognitive processes that come already kitted out in metaphors. Consolidation, which sounds like a word from management consulting, is more elegantly described as learning, or even enlightenment.

With Spritz, as usual, I was happy for the haters. Especially the ones with sciencey language and odd "centers" awash in public-private funds. Having written for an eternity about television—art's despised caste—I have an entrenched habit of ignoring what the social sciences say about linchpin cultural practices.

For a hundred years American studies have leveraged the word *science*—in hard science, soft science, pseudoscience, social science, neuroscience, Christian Science, Scientology—to prove that culture is bad for us. That's movies, games, theater, poetry, prose, television, music. Not a year goes by without new, lavishly funded studies, backed by universities or drug companies, that purport to detect pathogens in art and entertainment.

For this bad habit among would-be scientists, I blame Sir Percivall Pott. Pott was an English surgeon who became one of the world's first orthopedists in 1768, when he published a gripping report on the virtues of splinting his own broken leg instead of amputating it. But that wasn't Pott's only breakthrough. No: by connecting cases of scrotal cancer in chimney sweeps to carcinogens in coal tar in 1775, Pott showed that cancer can be caused by agents in the environment. This was a monumentally useful finding, made well before Louis Pasteur's work tightened the case for germ theory by connecting microorganisms and puerperal fever. Pott's findings put in place progressive labor laws governing the work of chimney sweeps, who were grievously underage and brutally mistreated; they also changed irrevocably the way we think about disease.

So how is Pott, a medical laureate of the first rank, to blame for the bad but tantalizing science that, year in and year out, makes cultural objects unlikely incubators of brain tissue–eating disease? Because Pott created a riveting literary genre: the "study" whose language conjures a causal, chemical relation between invisible objects in the outside world and the corruption of body and mind.

As an empirical argument about coal dust and cancer, Pott's case is irrefutable. As a polemic it is something better still: surpassingly satisfying and effective. Thanks to Pott, practices were banned, laws were passed, lives were saved. Who in any field wouldn't be inspired by the ingenuity of Pott, onetime apprentice in the Worshipful Company of Barbers—the haircutting monks who weren't allowed to shed blood and thus partnered with irreligious cutters from the laity, namely surgeons—who helped create orthopedics and then saved lives by detecting the carcinogen?

Alas, in the rush to try Pott's genre and borrow his infectious rhetoric, subsequent arguments in the same style dispensed with his actual analysis of scrotal sores and the composition of tar. Instead his landmark discoveries inspired nonscientists to purport (absent material evidence) to have found toxins whenever they wanted to crusade against anything they found morally objectionable.

Consider novels. Novel reading is today the defining practice of the sound and literate mind. If you don't read "for pleasure," after all—reading, that is, "not for work or school," in the words of the National Endowment for the Arts—you are said not to read at all; to read in the noblest and most salubrious sense in our time is to read novels. But in a rousing and beloved 1797 screed called "Novel Reading, a Cause of Female Depravity," novels are depicted as more sickening than coal dust in a boy's scrotum. And indeed they're sickening by the same dynamic. In impelling women to have premarital and extramarital sex, novels were not said to fill pretty heads with sordid notions (also a rank metaphor) but rather *to pollute the bloodstream*: "Without this poison instilled, as it were, into the blood, females in ordinary life would never have been so much the slaves of vice."

The "poison," in the anonymous ranter's telling, was not something that could be pointed to—not a molecule or speck of dust—but rather the voluptuous language and unsparing depiction of romantic passion that, it was argued, suggested to female readers that they ought to surrender their chastity and sleep with their friends' husbands. By sidelining actual evidence for poison while at the same time borrowing Pott's model of pathology, the ranter let his lively silliness slide ("as it were") into medical-sounding fact. Thus emerges the sheen of social science on garden-variety sermonizing.

It's anybody's guess why pathologizing culture is a cottage industry in the United States as nowhere else. A strong possibility is that art and culture are believed to cut into productivity. In a nation preoccupied with cradle-to-grave work and metastatic economic growth, productivity is naturally imagined to be a perfect synonym for health. Jeremiads against novel reading two hundred years ago may now seem ludicrous, but the culture-as-pathology genre, these "studies" bolstered by beetle-browed language that sounds so much like science it's called science even though no equations or invocations of natural law corrupt it, is having a heyday. Every year another book with a title like *The Shallows* or *The Dumbest Generation* cites these studies and condemns the Internet with no less righteous indignation than our Tory pamphleteer.

Exposure to Michel Foucault and historians of science at a tender age left me with no doubt that the language and apparatus of science are regularly deployed in a power play. This ought to be a truism by now, but still we need constant reminders. In the twentieth century unruly players like gay people were criminalized and called diseased under the sign of science. In this century writers with degrees in anything from social work to English cite "neuroscience"—no longer the biology of nerves but rather a vocabulary broadly applied to consciousness and philosophy and founded on the pleasing if theological-sounding 1950s theory of the triune brain—to regulate the magic of human experience. This includes memory, doubt, faith, focus, wonder, shame, sadness, ecstasy, tranquility, revelation. Their claims have a certain nervousness attached to them. Possibly this is because their primary tool (functional magnetic resonance imaging, the fMRI, which impressionistically represents brain activity) is notoriously unreliable. Worse still, for those of us interested in phenomenology, the claims of neuroscience rarely enrich the human experience they set out to elucidate. In fact they flatten it. They make it unrecognizable.

In this idiom the rapture of reading and the Internet become confounding sophistry and grandstanding—and the assertion, over and over and with mounting hysteria, that it's all a "cause of depravity."

Blurring the line between hard, empirical science and speculative social science, many neuroscientists cover up the inconsistencies of their method by pressing it into the service of venerable systems of value. If neuroscience, often but not always headquartered at Big Pharma, shows that massively multiplayer online role-playing games do something or other to the cerebral cortex, then the disgust many Establishment adults feel at a West Coast or Asian aesthetic—shapes, colors, sounds—gains satisfying moral heft. Class distinctions are shored up this way.

As an East Coaster myself, with a secular humanist background, canonical neuroscience often tells me what I want to hear: that reading is morally superior to video games, that face-to-face social life is hearty and true, that social media is empty and desperate. But it's still impossible to overlook its flawed methodology and transparent belief systems. As Elie Wiesel put it, "May I never use my reason against truth." False empiricism in the service of ideology will always show itself.

RHAPSODY

The bad reviews of Spritz only deepened my enjoyment of it. I reveled in the game-like demo. Eighty percent of traditional reading, claim the Spritz guys, is the laborious dragging of one's eyes across a line of text, and then down to the next line, like a typewriter carriage. You do none of that with Spritz. You just gaze and passively feast. By allowing galloping speed-reading, Spritz turned up a set of notes that had been low for me in the reading experience. It made reading somehow very direct. If ordinary reading was awkward for you, if you were or wondered if you were dyslexic, the process felt smoother. If reading already felt good, Spritz made it feel even better.

The words on the small Spritz screen, one at a brisk time, shot straight to my brain's reading spot, somewhere behind the eyes, perhaps in the region represented in Hindu art as the third eye. Working a gentle vibration on that esoteric organ of perception, Spritz turned it on without clobbering it. Words continued to amuse that place as my eyes and the rest of my brain seemed to take it easy. I didn't have to do anything but not blink, which didn't occur to me anyway. And thus, with Spritz, I read.

This new kind of reading was a revelation, and it shed light—as each digital development in reading and writing does—on the glorious mysteries of literacy. Probing my own experience, I discovered I generally find happiness in book reading when it's desultory, unregulated, and somehow truant. I confess to taking gross liberties with traditional books, savoring the rule breaking, skipping forewords, concordances, and boring chapters, while lavishing prurient attention on jacket copy, dedications, and acknowledgments.

On a single page of print or ebook I might savor the strangeness of a word like *batten* or *unlistenable*, or, less generously, I might fixate on a writer's tic and meanly tally how often she succumbs to it. While pursuing this private and pointless pleasure, I turn my back entirely on the SAT imperative of "comprehension." This desultory skimming is reading just as Roland Barthes argued we do and ought to do in *The Pleasure of the Text*, the short, eccentric, and moving 1973 booklet that presciently made an argument for how hypertext and digital content would be read.

In this way I consider reading to be reverie—and profoundly selfish. It's play and only play. What makes reading, beyond baseline competence, a moral virtue, to be tirelessly promoted by teachers, parents, and first ladies? I can't fathom. The way my children and my mother and I read it's closer to a bad habit, like binge-eating or sleeping all day.

The surprise of Spritz is that, even as it makes skimming and skipping around impossible, it suits me. It's a new game, bearing only a family resemblance to reading. I embraced Spritz the way I embraced app games like Angry Birds and Hundreds. Spritz takes less decision making and backing-and-forthing than does regular reading, and more stillness. Possibly more passivity. Maybe rapture, or stupor.

In any case words keep coming, and to my delight my mind

keeps taking them in. At the same time Spritz makes the analysis of symbols more game-like by incentivizing speed. You can set a words-per-minute rate with Spritz and see how well you keep up. When I faltered at 400 wpm, I felt enjoyably challenged to try harder. It felt like pedaling a bike uphill on a low gear, like trying to keep up with revolutions of a wheel.

"Not really reading," the *Atlantic* nonetheless concluded, *ex cathedra*.

GOOD READING, BAD READING

This wasn't the first time in the digital revolution I'd heard about good and bad reading, real and false. Years earlier ebooks were said to be illegitimate because not redolent of mold or hospitable to marginalia. Before that, email was said to lack the intimacy or warmth or manuality of authentic epistolary communication. Texts and tweets are still being dismissed as too short to be important. Occasionally someone says, *Oh but haiku can be beautiful, and it's short*. Never mind that the *majority* of the world's great epigrams, aphorisms, and *pensées*—and certainly anything composed in an ideographic language, like Confucianisms—are handily expressed in 140 characters, often with many to spare.

You see how ensnarled the issue gets once you call reading true and false. Before Web 2.0, in 2002, when anxiety about digitization was especially bald, the National Endowment for the Arts asked seventeen thousand unsuspecting Americans whether during the previous twelve months they had read any novels, short stories, plays, or poetry in their leisure time, not for work or school. Two years later the NEA returned its shocked report. "Less than half of the adult American population now reads literature," it said, predicting an imminent Dark Ages. "As more Americans lose this capability, our nation becomes less informed, active and independent-minded."

What a question. Have you read any novels, short stories, plays, or poetry *in your leisure time*?

I couldn't help but think: Leisure time? Few of us are swimming in leisure time, and those who are probably don't use that time for reading. Rather we read for work or school. On-the-job literacy, in fact, is consistent with the American ethos that made everything from *Poor Richard's Almanack* to *The Lean Startup* into best sellers. In a country suspicious of leisure, we often tell ourselves that we read "for work or school"—that is, to fulfill obligations and gain instruction. Even voracious readers of mass-produced Harlequin fiction, according to Janice Radway's landmark 1987 study, *Reading the Romance*, regularly praise the books they read as educational rather than escapist. Aspirational love stories double as seminars in subjects like wine, furs, tiramisu, and Tuscany.

Americans read with highlighters. We read for "information," as though for a future comprehension test. We underline, copy quotations, pull excerpts, produce decks, compose reviews. And if the World Wide Web has shown us anything at all, they also *comment like crazy*—on literary blogs, on Facebook, and on Goodreads as well as in seamier venues.

This reading-writing or participatory reading has come to be regarded as impure reading, however. Pure reading, as the NEA has it, is for a kind of arts-besotted *belle époque* figure awash in leisure time. Turn to sonnets when you're sad or in love, not to get an epigraph for a final paper or a reading for a co-worker's wedding. That kind of avocational reading is certainly possible—50 percent of respondents told the NEA they read that way—but it also seems clear that reading for work should count as reading too. Codices, scrolls, leisure, work, epic poetry, tweets: let's call it all real reading.

READING IS LURKING

What's especially strange about the way literary institutions enshrine leisure reading is that exactly the opposite goes on online. This is yet another way that the digital revolution, like any revolution, neatly inverted the cultural values that preceded it. I found this out in my many years "lurking" on websites devoted variously to parenting, consumer electronics, celebrity gossip, furniture design, health anomalies, and real estate. To the sites' message boards, which I used to follow avidly, I contributed a total of three overwritten comments. They sank like stones.

I was, for the most part, at peace with that. But every now and then, when the posters on these sites muttered about "lurkers," I'd shudder like a Soviet mole in the Pentagon. Because, yes, I lurked: I visited boards but didn't post; I took without giving.

If lurkers are condemned with less venom than other message board scourges—spammers (who post ads), trolls (who promote controversy), and sock puppets (who burnish their own reputations under aliases)—this may be only because lurkers are ubiquitous. Everyone, even lightning-fingered texters who never stop composing, has to simply read sometimes, even if only to take in responses to her many texts or get an address off Google Maps.

That said, lurking defies the spirit of interactive media. Years ago Michael Hirschorn praised Facebook in an article in the *Atlantic*: "You have to give information to get information." He admired, in essence, the tariffs Facebook levies on lurking. Online you're expected to interact, not hide, eavesdrop, and slink away. But what name did it used to go by, this practice of anonymously sitting back and taking in long sequences of words without producing any yourself? Hey, wasn't it once called "reading"?

In short, in my heavy lurking days (which ended as my confidence in the voice of my online avatar grew and I switched to the less-readable mobile Web) I was caught between analog and digital systems of value. Was lurking a devious violation of Web ethics or a return to luxurious nonparticipatory reading? It still feels indulgent just to sit by and, to coin a phrase, let words flow into my brain. When I lurk, I relax, fall silent, become a cosseted baroness whose electronic servants bring her funny pictures and distracting tales. I have no responsibilities. I'm entirely on intake. If I were reading Knausgaard or Anita Shreve this way, I'd be an NEA-certified exemplar of civilization.

SYNESTHESIA

One winter I read perfume blogs. Nights I would lose myself in a sybaritic rapture, the clock on my screen racing toward morning as I played a bait-and-switch game with my senses, taking in words that describe smells by calling on images, textures, music.

For those cold months the sensory pileup felt good. The blogs told me that scents ring, sing, and lash out in fury; they also cradle, buffer, withhold. I believe them. Too often personal computing is framed as a sensory strain, a source of repetitive stress, eye fatigue, and sciatica. But that winter at the screen came to feel invigorating, even indulgent. The screen smelled only of the flammable solvent I use to clean it, but thanks to the synesthetic stylings of the perfume blogs—one matches scents with paintings; others evoke scents with nostalgic vignettes—it heaved with olfactory delights.

But these were uncertain, ambiguous delights: perfume inspires writers to mad flights of overwriting, extravagant hymns, and purple poetry that supply almost no hard facts. Chandler Burr, the perfume critic, reminded me that even perfume's traditional musical conceit—in which clusters of raw materials with names like ozone, neroli, and blue musk are called "notes"—explains nothing at all about a scent's artistry or precise chemical composition. (By email he urged me to think instead about that screen-solvent smell, which I had taken for mere ethyl alcohol: "All these things are scented.")

But come on, who cares about chemical composition when it's two in the morning! Chanel No. 19, according to a review that Burr often quotes, maintains "unripe greenness like a tense unresolved musical chord to the very end, without succumbing to sweetness." En Passant, by the lights of a sniffer on the site Now Smell This, is "sunny and cloudy at the same time, full of the enchantment of the forthcoming storm or the everclear sunshine." A Pink Sugar detractor on another site describes the scent as "cotton candy that has caught on fire and sizzled down to a black gooey mess."

What a world. I couldn't get enough of this prose syrup, and yet to me such voluptuous descriptions didn't actually evoke green fruit or incinerated candy. Instead they inspired more attenuated fantasies about acquiring perfume, holding in my hand a substance that could evoke storms, sunshine clouds, or smoldering black candy goo. Let's face it: I was *reading* about *smelling* something that might remind me of life. If there's a greater remove I could have achieved from visceral human experience, I don't know it. Or maybe I do know it. As a solitary, late-night pastime, done by artificial light, this reverie came prepathologized, always already shameful. In that way perfume fantasizing while at the mercy of words is, like lurking, reading.

COLD HUNTING

One cold October morning several years ago I found myself gazing out at the tarmac of Kennedy Airport. At my side was a factory-made pastry in a waxy sack from Au Bon Pain. My plane was delayed. I was privately overjoyed.

A wait at the gate, and then a flight to California. As ordinary as it gets, and yet: nearly six airborne hours! I anticipated the enforced freedom from conversation with excitement bordering on euphoria. If you're lucky and the wifi is broken (a rare pleasure lately), you can still feel the old Major Tom factor to air travel now: silent go the devices as up we rise, while the taut invisible Web wires snap one by one until finally we're floating in a placid immaculate zone.

Even when wonky and expensive in-flight wifi is on offer, airborne tweeting and texting come with more than the usual friction, and the altitude offers a welcome brake on connectivity. It's stolen peace. If the airlines knew how precious that icy aloofness was to some passengers, they'd find a way to make us pay for it. The JetBlue ColdSpot.

For that particular flight to Los Angeles I was almost erotically giddy because I was taking a Kindle traveling with me for the first time. This was the first-generation Kindle, released in November 2006 for $400, at which price it sold out in six hours. Ideal for reading novels by Marilynne Robinson and Ian McEwan, my Kindle had rapidly became something more: a refuge. An escape from the speedier Internet I hadn't even known I craved. But the design and user experience of the Kindle—yes, that's UX, the snowboarder-style abbreviation that refers to everything from the look of a screen to an app's or gadget's navigation tools—now spoke to me of blessed sequester and holy retreat.

A gadget tied to Amazon that offered retreat from the Internet? It was simple: that first Kindle, unlike the other devices I carried, didn't pulse with clocks, blaze with video, or squall with incoming bulletins and demands. Instead it was reserved, with something on a continuum with a stone's integrity and self-reliance. Exiled from the sparkling crisscross of fiber optics and infinite cybernet works, the Kindle was almost like a book. Thus the Kindle protests the Internet, as the App Store protests the web. Technology in productive tension with other technology has a special pleasure to it.

My reflex for reading long, involving books—dampened by age, I used to believe—came bounding back the moment I saw the Kindle. I instantly forgot how uneasy I'd become around books in my thirties. As the Internet increasingly captured my attention, books had come to seem not like a sanctuary but like a dungeon; while reading them I feared missing out. And on what? Simple: on other kinds of reading. Websites. Blogs. Tumblr. While it's commonplace to worry that the rise of digital culture, particularly the Web, endangered the practice of reading, the Internet has actually tended to hurt only one *kind* of reading, which happens to be the kind long identified with intelligence and good taste: book reading.

As all this hierarchizing of reading makes clear, how a person makes meaning of written language is a significant marker of her identity. It's also a contentious one. In the current thinking about written communication, reading and writing SMS notes—texting—is considered deeply pleasurable, addictive, subliterate, distracting, and dangerous. By contrast, reading reputable-looking thick, bound books on paper—on virtually any subject at any level of syntax, organization, and diction—is considered exceptionally challenging, uncomfortable, and noble.

Thus by the millennium's second decade digitized words—good ones, bad ones, hurtful and benign ones—are rarely out of our field of vision, and all anyone seems to want to do is read them. Reading has become so ubiquitous and compulsive that now it's reading while driving that endangers motorists more even than intoxicated driving. Or maybe the reader and the inebriate are at last one and the same. (Baudelaire: "Be drunk always. . . . But on what? Wine, virtue or poetry, as you wish.") As Michael Pollan memorably chronicled in his book *The Omnivore's Dilemma*, America in the age of rye surplus used to be a nation of drunks, then the overproduction of corn turned us into overeaters. I'd add that, as words have proliferated hypertrophically on the Internet, we've become a population of overreaders, of hyperlexics.

In 2013 Jeff Bezos, founder of Amazon, father of the Kindle, acquired the *Washington Post* and proposed that digital "customers" now be called "readers." We citizens of the Internet compulsively register and interpret symbols. As a result symbolic analysis—not agriculture, industrial labor, finance, travel, commerce, or consumption—is the primary activity of this civilization.

As strange a turn of events as this may seem to those who imagine that cultures coarsen over time, where coarsening is understood as moving from more literacy to less, this development really should surprise no one. Unlike physical sensations and sounds, language is *already* a symbolic affair. It didn't take digitization to turn it into zeros and ones. Braille and Morse code handily made language binary. Black ink on white paper: presence and absence. It might be aesthetically controversial to digitize music and photography, with their spectra, scales, gradations, burns, dodges, sharps, and flats. But language and words are self-digitizing, were digital before digitization.

Digitizing written language—in short communiqués, as in texts or tweets or the speed-reading samples offered by Spritz, or in long form, as in lengthy Facebook debates or Amazon's ebooks—even restores something to written language that print technology had cost it. Let's call it mentalness, proximity to pure thought. When changed from chunky, chewy, smelly matter (think used bookstores) to the photons that enliven our screens, written language becomes purer somehow, and more itself.

Digitization also makes the written word proliferate and circulate at dizzying speeds. It turns those words headier, faster, and trickier. Like thoughts.

Or that's what I discovered with ebooks, anyway. I had become bored with the books on my shelf, the wide tables of deeply discounted and interchangeable best sellers at Barnes & Noble, the indifferent clerks at the major bookstores and the embittered clerks at the indie ones. The tatty galleys I got to review and even the promise of enlightenment in the deep, dark stacks at Widener or the Bodleian: all of it had started to conjure atrophy, morgues. Death itself.

For people who genuinely like to read—to replace their own trains of thought with other idioms and language streams; to surrender, for a time, their own internal symbolic order and come alive to a foreign one—an ebook is heaven. No, an ebook is never going to suit those bibliophiles who cherish the scent of bindings and the material resonance of printed artifacts in time and space. But not all bibliophiles are readers. ("The non-reading of books," writes Walter Benjamin, is in fact "characteristic of the collector.")

At the same time, as the rise of ebooks has demonstrated, the vast majority of readers are agnostic about actual three-dimensional books. We love to read—BuzzFeed lists, Snapchat captions, weight warnings in elevators—but find chunky codices often demand too much attention to their material selves. That design! That typeface! That heft! I'd rather all that clamor by the material world backed off and left me alone to read.

I'm far from alone in preferring reading to books. Ebooks have outsold print books at Amazon since 2011. By 2017 the consumer ebook market in the United States is expected to be bigger than the entire consumer print book market. Print books, which date from Gutenberg in the 1450s and which millions of machines and humans and warehouses and infrastructure exist to produce, are losing to ebooks, which have only been around a decade or so.

There would be no ebook penetration at all without the peerless Kindle and its merely serviceable design. Amazon's ereader is plenty user-friendly; it just isn't user-obsequious. It doesn't creepily mine you for data, like an ogling lech. It doesn't read *you*, as the Web does. (On the Web, "if you're not paying for it, you're the product," as the saying goes.)

Blessedly you are not the product on the Kindle. You pay a hundred or so for the awkward device; you pay again to load it with books. Now the plain Kindle sits patiently while you read *it*. To read on a Kindle you have to go, mentally, more than halfway to the experience. You have to concentrate, integrate new material, persevere when you might be stopped by thorny words or elusive concepts. Kindle words don't light up or move or turn into cartoons. You are alone in making meaning of them, and that's a reader's cherished duty and her deepest pleasure.

Unlike with a tightly designed Wii game or binge-worthy Netflix show, Amazon's Kindle also keeps the parameters for reader frustration ("This is too hard!") and reader boredom ("This is too easy!") relatively broad and true to life. If the concepts or style of a book baffles you, nothing in the Kindle design condescends to soothe your feelings. If your attention wanders during a dull passage, by contrast, there are no videos to divert you. A sustained encounter with a book on the Kindle is a suitably demanding cerebral event.

And thus the Kindle restored to reading a nineteenth-century rhythm and, increasingly, nineteenth- and even eighteenth-century prosody. In the reader's experience this rhythm is associated with the heyday of middle-class reading, when reading long books—sentimental fiction, serialized novels, embellished memoirs, religious tracts—was so pleasurable it was considered, as in the "Female Depravity" argument, a hazard to the brain. (An art form or cultural practice thrives to the degree it is considered poisonous; by contrast, it's ailing when there are MFA programs in it.)

That was the era before design obsessives of the twentieth century made fetish objects of opulently published books and began to promote bibliophilia and graphic design over literacy. It was certainly well before anyone in the twenty-first century had even seen the blinding backlit screens that turned everyone's field of vision into a personal Times Square or Shibuya.

The Kindle screen is a throwback. Its first low-energy "electronic paper" looked dusty, like Victorian newsprint. It resembled neither an art book to be found on the table of an immaculate living room, nor a Scribner showpiece with groovy typeface, nor an iPad. As a piece of consumer electronics, the first Kindle was a joke. Its bumpable buttons constantly flipped pages. Its pointy keys were stiff and useless. Its homely lily-white casing turned dingy after one book. Reviewers found the design easy to dismiss. On Amazon's own site reviewers called it "a disaster." Macs and iPods had trained consumers in what gadgets should be: high-design commodities of such extreme tactile pleasure that users used to report a desire to lick them, copulate with them.

No such urge possessed early adapters of the plain-Jane Kindle. Thus the Kindle was never treated as a gadget. It was not sensual or fetishistic. It was not a vanity device, supercharged with processing power and razzle-dazzle for male connoisseurs of low-mass, hi-fi, high-speed, high-def sounds and sights. Just as PBS and then HBO styled themselves as TV for people who shun TV, the Kindle was designed as tech for people who disdain tech—and certainly disdain the rambunctious and irritating World Wide Web of 2007, with its ideological blogger posturing, maddening hyperlinks, and balky video. In the 2000s these headaches were grimly believed to suggest the future of reading.

The Kindle was a gadget for readers of long works. A gadget, then, disproportionately for women. (In an AP survey of "avid readers," women read nine books a year, while men read five.) And as any retailer knows, that's where the money is. In 2008 Oprah Winfrey, standing with Bezos, hawked the Kindle as "life-changing" and "the wave of the future." In a nod to that year's brutal economy, Winfrey admitted that the Kindle was "expensive in these times, but . . . not frivolous because . . . the books are much cheaper, and you're saving paper." Here was a frugal, environmentally sensitive purchase for women with responsibilities to home economics but also to the Oprah-promoted "me time" or "self-care," with its requisite scented candles and stack of novels, inspirational literature, and self-help. The books on Oprah's first Kindle, she told her fans, included *The Story of Edgar Sawtelle* by David Wroblewski, *The Audacity of Hope* by Barack Obama, *Ageless: The Naked Truth about Bioidentical Hormones* by Suzanne Somers, *The Alchemist* by Paulo Coelho, *The Forever War* by Dexter Filkins, and *Crack the Fat-Loss Code: Outsmart Your Metabolism and Conquer the Diet Plateau* by Wendy Chant.

For five years now Amazon has sold millions more digital books than traditional books. At the same time, the world's biggest dry-goods concern was now selling more *books* than ever before—including, for the first time, print books published exclusively by Amazon, without agents, publishing houses, or outside distributors.

WHO WILL PAY $40,000 FOR A BOOK?

Self-publishing has been an extremely important phenomenon in the evolution of digital language. It has also been a wonder to watch. According to Bowker, which uses conservative figures because it tracks only ISBNs (which many self-published books don't have), there were nearly 550,000 self-published titles in 2015.

I discovered some early consequences of this turn of events when, several years ago, I gave an address to a paper trade association at the opulent Drake Hotel in Chicago. Its organizer had let me know in no uncertain terms that the group, charged with pricing paper and pulp while accounting for variables like blights to arboriculture by clearwing moths, was looking for bullish signs for the paper business. Clearly it had been hit hard by the rise of digital and the decline of paper newspapers. Could I offer the papyrophilic audience in Chicago any words of reassurance? Was paper making any kind of comeback?

Obligingly I searched for bright paper news—and found little. The Web and then ereaders (including, by that time, the Kindle app for iPad and iPhone) had been a world-historical boon for readers but had left paper, books, bindings, typefaces, ink, warehouses, publishing houses, designers, packagers, egos, and all of the agents and sales folk attached to printed words on paper in stunned sorrow. The runic proofreading marks I had studied as a fact-checker at the *New Yorker* and practiced making on big paper galleys with No. 2 pencils at my desk were nearly obsolete.

By contrast, if you knew even a touch of simpleton HTML, you could bill yourself as a genius. Magazines were shrinking, spluttering, then folding. Newspapers too. Publishing houses were laying off everybody. From the foundries, I could have told the finance types in Chicago that publishing in New York was Allentown: they were closing down all the factories.

Instead I told the sharp-dressed crowd at the Drake that toilet paper sales seemed to be holding steady. Cellulose, the tree matter that bulks up McFlurries and other food products, was also doing just fine. But what really piqued the crowd's interest (gentlemen scribbling in notebooks at this point and one furtively tapping an iPad) was some news I delivered about self-publishing. To wit: even if demand for printed matter was down, there were *some* people willing to pay for beautifully produced books—and handsomely. And they wouldn't pay just $10 (then the Kindle book price) or $30 (the typical hardcover print price). For the right book they'd pay, oh, say $40,000.

These high rollers were authors themselves. As Walter Benjamin put it, the best way to acquire a book is to write it yourself. It's also the fanciest way. Many writers—myself included, if I'm honest—are perfectly happy to read ebooks and don't care if the whole Western canon goes digital, but we want *our own books*

Moroccan-bound on good paper in a nice Garamond typeface, engraved, ideally. And naturally produced in vast, vast quantities and then lovingly shelved in venerable libraries. Funnily enough, self-publishers (authors) seem willing to pay for all this—to hire editors and designers and publishers to produce print runs of the books they write. An analyst at the Drake followed me at the podium with hard numbers backing this up, and there was some relief among the pulp-and-paper salesmen.

What I witnessed that day was a moment when the interests and fortunes of the paper people sharply diverged from the interests and fortunes of readers. Bibliophilia, as I've been saying, is not the same thing as literacy. Writers like printed paper. Value to them inheres in the commodity that is paper, made more valuable still when inscribed with words and thereby coined like a gold ingot. (In 2015, when a group of us who'd been writing for Medium decided to have Blurb print a few copies of our first year of work—at the suggestion of Craig Mod, a great proponent of print—everyone jumped at the idea. Including me.)

Many readers, by contrast, find value in light, versatility, portability, photons, and pixels—and abstraction and distraction rather than solidity and matter. Neither of those interests is more noble than the other; compulsively reading mass-produced diet books and pseudoscience (as Oprah and I do) seems morally about on par with selling paper, if far less lucrative. But certain values and shibboleths and centers of profit and meaning in the American middle class are potent shapers of identity, and I saw them coming apart that day in Chicago.

MONOMANIA

Though I'm often as happy to read a short article on Medium as a novel, I'm not by nature a hummingbird. I like to be able to pursue monomaniacal reading obsessions. One myth about on-line reading is that it's all skimming and partial and no one reads in a sustained, focused way anymore. Indignant observers have said this about culture since the birth of the codex.

True, readers might not read the way writers and publishers have laboriously mapped for them, starting with "Call me Ishmael" and ending with "only found another orphan." But watch someone with a laptop determined to put together a biography of Lou Reed for himself and you'll see what focus and determination look like. He goes to Wikipedia, then tracks down links, then listens to a song or two on Spotify, consults a lyrics sheet online, then looks up Nico and Laurie Anderson, then reads the *Times* obituary—but maybe has to pay to read it, having used up his free monthly articles, and thereby gets involved in entering his credit card information and more—and finally ends his search by reading twenty-four reviews and ordering an ebook biography of Reed from Amazon. Maybe at some point he posts on Facebook or Twitter to get recommendations. The whole research episode lasts ninety minutes, and he's pulled together a decent Lou Reed monograph, tailored to exactly what he wants to know.

That's called "surfing" or sometimes "stalking," and it may appear unfocused. In fact, as the lexicographer and publisher

Lizzie Skurnick puts it, it requires no less than *rapt* attention not to lose one's place. Reading a book straight through is much easier. For many of us this kind of Web project is also highly satisfying.

What do I like best about the Kindle? Easy: the Internet connection. The so-called Whispernet is so poor, its browser so ungracious and inaccessible, that you're discouraged from ever exploiting it except to order books from Amazon. And this weak Internet connection and elusive browser turned out to be the most savory parts of the machine. It kept the Kindle insular and remote from buzzing data swarms, but it also made the Kindle something more than a book. Because while I might like a few hours on an airplane, I don't want to move into a locked library carrel and never visit the Internet again.

So the Kindle, like a good reader, is immersed and absorbed but not completely out of it. The first Kindle acknowledged the Internet. It heard its boisterous demands; it just ignored them. This means the Kindle bestows on the contemporary reader the ultimate grace: it keeps the Internet at bay.

Download a good book—a novel by Elena Ferrante, say— and the design flaws of an ereader become irrelevant. The device circumvents the fussy design critic in us and finds instead the serene and long-neglected reader. With a gray screen that uses actual black ink that has been given an electric charge (wow), the Kindle means you can lose hours to reading novels in one sitting. And yet the slight connection to the Web still permits the sense that if the apocalypse came while you were shut away somewhere reading, the machine would get the news from Amazon.com and find a way to let you know. Anything shy of that, though, the Kindle would hold your calls. It would leave you alone.

It's bliss. Emerge from the subway or alight from a flight, and the Kindle has no news for you. It's ready only to be read. It's like a good exercise machine that mysteriously incentivizes the pursuit of muscle pain while still making you feel cared for. The Kindle makes you want to read, and read hard, and read copiously. It eventually also makes you aware that, compared with reading a lush, inky book, checking email is boring, workaday, and lame.

But no sooner on that 2007 flight had I decided that, for the moment, I'd discovered in the Kindle a way to tame the anxiety of the demanding digital world without abjuring its pleasures, when I found myself explaining the device to my seat mate on the plane. He asked! As I babbled on about it, I suddenly realized that the Kindle is, above all, uncool. This man furrowed his brow as I praised the Kindle's uneasy relationship with the Internet. He looked at the gray screen and said, with authority, "That's way too dim." He pulled out his brand-new iPhone—a 3G, the state of the art at the time.

To my discomfort, I struggled to return to my novel, but after ten minutes of self-consciously reading and rereading the same pages, I got into it again. The shadowy hue of the page and the letters of digital ink became my whole world once more. And my seat mate, besotted with his phone, had nothing more to say to me. My Kindle announced me as an oddball, a wallflower. A reader, then.

MISTAKES ARE STYLE

But *what* are we reading on all these devices, and how are the devices and applications influencing the language on them? Twisty narrative, for one. It's no surprise that the Harry Potter generation, who grew up on serial fantasy fiction set in the labyrinthine hallways of a magic school—as well as on suspenseful literary television like *Lost*, which is built on cliffhangers, errors, and indirection—have taken to the disorienting pagination of the Kindle and the lost-in-the-funhouse narrative that thrives in ereader form.

These are books that defy the miserable school essay format: tell them what you're going to tell them, tell them, tell them what you told them. They are filled with fake-outs and unfulfilled promises. The best Kindle books serialize easily and pulse with suspense that drives the reader emotionally (rather than rationally) through the maze. There are mistakes, redundancies, false starts, red herrings, loose and dead ends. *Purity. The Goldfinch. Fifty Shades of Gray. Gone Girl.*

Off the ereader, on the Web, another kind of prose has formed itself, also characterized by florid mistakes. The hacker world is still studded with hoary in-jokes surrounding typos like "teh," which gives a certain brand of ham-handed typing a recondite meaning all its own. "Pwned," a notable example, signals humiliation of a rival, as in mistyped "owned," or the deliberate compromise of another's security system. Initially I thought the purposeful typos were meant to stymie search; someone looking over 4chan for the culprit in a security breach might search for "owned" but not "pwned." Then I thought that the typos made fun of the speaker as klutzy and dumb. But now I think *teh* and *pwned* are just silliness crusted over by tradition, like Cockney rhyming slang. In any case, the embrace of mistakes: that's a hallmark of Internet prose.

This flexible, mutable style with its errors flouts the "lapidary" style, coveted in the twentieth century and apotheosized in print magazines like the *New Yorker*, with its famous commitment to "unerring" use of semicolons and em dashes, especially in the poetry. (Remember when a poet was someone who could work for days on how to place a comma?) What we did at Yahoo! News, which the staff called "retroediting," would make the *New Yorker* staff blanch: we'd post something as soon as the sloppiest draft was ready and edit it *after* it was available to readers.

Translating Web language for print, which I did for years in aiming to make digital culture legible in the *Times*, is not easy. I remember wanting to excerpt a passage signed by one "zipthwung," an astute online commenter: "Pornography if for the ruling classes and their violent vulgar all consuming appetites. Or their slaves."

I was at a loss: Should I repunctuate this, adding commas and plunking a hyphen into "all-consuming"? Should I turn that "if" to "is"? Maybe zipthwung *meant* "if." These kinds of questions still dog me when I live between analog and digital forms. Sometimes I opt to copy words and paste them into the text of a column, to quote verbatim. I'll treat message-board words or words on Facebook as if they had been written in books, articles, brochures, or press releases. But is that what the writers want? Should I care?

Consider another example. To show that Web users are curious about human reproduction, I might quote the great Kavya on Yahoo! Answers, word for word: "How is babby formed? How girl get pragnent?" But that makes Kavya look like an idiot. Readers might miss the sweet earnestness of his question. Maybe he (or she) is seven or a native speaker of Hungarian. I should cut the kid a typographical break; that's not an easy question to ask. The cockamamie diction and syntax of Internet English is, possibly, only incidental to his inquiry. A reporter could paraphrase or revise his question—"How is a baby formed?"—lest readers are blinded to the intent of the question by moronizing typos.

But "How is babby formed?" is funny. And who wants to deny readers a chance to laugh and to get the full flavor of Internet culture wackiness? It's flat-out lying to pretend that everyone (or anyone) spells well online.

My problem with message-board language brings up a prior problem in journalism: the difficulty of translating spoken language into written language. Jacques Derrida gained notoriety by dimming the bright line between what was known in strange pre-Internet lingo (French, was it?) as *langue* and *parole*. He thought the written-spoken distinction was suspect and by turns collapsed and reasserted itself in the merry game of signification.

Nothing works more Frenchly and merrily this way, shape-shifting at a rapid pace, than Internet language, which morphs from standard English (a dialect of which has become the Web's lingua franca) to other languages and dialects to slang and emoticons and acronyms and phonetic miscellany. (Take "Truf. NM. #factfail." Can this tweet, in response to another tweet, be taken as an admission of some kind of error? Can it be faithfully paraphrased as "She admitted her mistake on Twitter"?) I can't tell how much of this key-cap casserole belongs in ink on paper or how much of it makes *sense* there.

The Sanhedrins of style at the last great daily broadsheets are not so amused by the Derridean game of signification. Most of them seem to believe in standardizing spoken English—to a point. At the *Times* using nonstandard spelling to reflect dialect—"He wuz a good friend"—is seen as a sketchy business, since no two writers do it the same way and since it can reflect bias. But rhetorical eccentricities ought to be preserved. "We're friends for twenty years," for example, does not have to become "We have been friends for twenty years."

Some architects of style have proposed that communication on social networks should be treated like the text of a novel. As novels of sorts, social media ought to be excerpted using the same protocols that critics use to quote fiction. That is, we should go light on the academic *sic*s, addition brackets, and omission ellipses, which in a paper can come across as sneering, cluttered, pretentious, or all three. By contrast, when transcribing posts, idiosyncrasies of language should be preserved as far as possible and taken as intentional, unless in context they are obviously evidence that the writer has innocently hit the wrong key ("hlelo," "rihgt"). A "wuz" on the Internet remains "wuz" in the paper. In thorny cases a critic or reporter can extenuate a passage outside of quotation marks. ("'The soiled fish,' writes Melville, conjuring an odd image with 'soiled' where perhaps 'coiled' was intended.")

Daniel Okrent, the first public editor for the *Times*, explained it to me this way: "The minute you start trying to replicate someone's accent or diction, you run the risk of appearing to be patronizing or worse. When the Mississippi State football coach said something like, 'There ain't but one color that matters here,' the paper was wrong to recast it as 'There is only one color . . .'—he didn't say that." Okrent continued, "But if in reaching for the sound of his voice they had rendered it as 'I ain't gonna suspend mah players fer actin' up on weekends,' it would have been inappropriate. I say stick with the actual words the man uses and not with the way he says them."

Dropping g's, Mark Twain–style, does look supremely corny, though newspapers once liberally clipped those g's into apostrophes for folksy effect. In 1907 the *Times* published an article titled "Mr. Devery Has Some Thoughts on the Way Things Is Goin'." Devery, a former New York City police chief, was what can only be called a colorful character, complete with colorful, g-free words and colorfully disagreeing subjects and verbs. "If things is run right," Devery opined, "the chief of police ought to be nothin' but a sort of foreman, a feller to carry out the orders of them above him. He ought to be a sort of—of—editor."

Comes off kind of fakey today, don't it? Certainly having one or two subjects say "goin' " or "gonna" or (come on) "gwine" when everyone else's participles get standardized is unfair and misleading. On the other hand, readers of a hundred years ago found Devery's dialect funny, and writers and readers alike crave funny quotations. It may seem condescending or even racist to use the dialect conventions of *Pudd'nhead Wilson*, but it also seems like a crime against humor and the truth of Web language to adjust "How is babby formed? How girl get pragnent?" in the name of imagined fairness.

Ultimately I stand by my consumer-end experience— my reading experience—of social media. Though an intrepid A-section reporter might be able to turn up names, ranks, and serial numbers by pushing sources and insisting everything be said on the record, in proper English, and for attribution, I'll never accomplish all that with the crazy stuff on boards. Certainly not in a passage like this one, which showed up recently on a mothering site:

> How many months into your relationships has ILY come
> out?
> 3ish
> What are you, 16?
> just curious

what is ily?

I love you

Idiot

And who would want all those names, professions, ages, locations, David Copperfield kind of crap? So I'll go on treating message boards like novels until I am persuaded otherwise. Oh, dear Web, I love you, Idiot.

HASHING

After windy storytelling and uncorrected mistakes, hashtags are a cornerstone of Internet style. An invention of the first tweeters in the early days of Twitter, phrases in hashtags tend to amplify, ironize, or otherwise contextualize a message in regular text. They also make a tweet searchable, as they are meant to mark the subject of a tweet. (Today they're also used on Facebook, in personal texts, in email, and even, by millennials, in speech.)

In learning to read—that is, to resist reading—on the Internet, you never know what story will break down your defenses and make you read it. Could be Israel, could be health care, could be *Girls* on HBO. Or it could be #freeskip.

For some reason the arrest several summers ago of Henry Louis Gates Jr., the Harvard professor, wouldn't let me go. The police reports weren't enough. The news accounts weren't enough. The White House intervention wasn't enough. The opinions of dozens of blogger-sages, including Gates himself, as well as writers like Stanley Fish and Christopher Hitchens, who liked to finish off a subject, haven't been enough.

This has happened to me many times, with the lost Malaysian plane, juvenilia by Hillary Clinton, and crime in Appalachia, so I knew what was up: I'd really never be satisfied. My attraction to the Gates arrest narrative, with its potential for curiosity, surprise, indignation, and pedantry on themes from race to police procedure, academia, and the history of Boston, struck me as a craving induced by industrial design, like southwestern egg rolls at Chili's. Not until the whole of Gatesgate had been unpacked, as people said in graduate school (where, full disclosure, Gates had been my adviser for a time), could I move on. Was anyone with me? Or did everyone else healthily revolve with the news cycle?

Enter #freeskip and hashtags. Hashtags, of course, are curious words and mashed-together phrases earmarked with a hash symbol (once known as the pound sign). They work sort of like the moment in a conversation when a big talker might say—generously to a newcomer, pointedly to a dummy—"We're talking about the future of the Democratic Party here." A hashtag (think #DEM2020) is also a link, so anyone who encounters one on Twitter can instantly search the network for that phrase.

Where library science uses shared, intuitive, and (in principle) value-neutral systems for organizing information, Twitter users often classify their tweets in the most condensed, most charged, and least transparent way possible. While aiming to draw people in, Twitter users nonetheless strive for unique hashtags (#freeskip instead of #gates, for example) so that searches don't retrieve off-topic stuff. I can say this from experience: if you urgently wanted to know about the Gates arrest, you wanted to dodge tweets about Bill Gates. That nonnews was for #billgates people.

As a subject crests in the popular imagination, certain hashtags become affixed to the story, or to a way of seeing the story. You tend to be loyal to your hashtags, at least for a few days, and track them wherever they go. Maybe hashtags are like those color-coded bandannas that gay men began sticking in their back pocket in the 1970s to show that they were into this or that kind of sex. Only way dorkier and not sexy.

I found #freeskip soon after I read the first news report. Looking for someone to compare notes with, I plugged "Henry Louis Gates" into the Twitter search box. A tweet by someone passing along an insight from @colsonwhitehead (Colson Whitehead, the novelist) led me to this: "One thing's for sure: This wouldn't have happened to Jean Toomer. #freeskip."

Ha! Perfect Twitter sally. I abandoned my more general sucker's search and clicked on "#freeskip." I specifically wanted to hear more Toomer references and read the remarks of people bold enough to call the august professor Skip; I wanted to join what I came to think of as the #freeskip *people*. Indeed it seemed to me that colsonwhitehead and others using #freeskip were serving up the best flash insights and asides. For days I stayed close to #freeskip, refreshing my search like a drugged monkey.

Hashtags are hard to explain partly because they are almost never transparent or ideologically neutral. They're meant to be code. Where Colson Whitehead might not use Gates's nickname in a CNN interview, @colsonwhitehead uses it with abandon as a tag. What's more, his hashtag evokes the irreverence of "Free Winona," the slogan that spoofed the seriousness of "Free Mumia" in 2002. That's what drew me to it: an Internet opiner must be pretty bold so early on to liken Gates, however indirectly, to both a former Black Panther on death row and a Hollywood shoplifter. Me, I wouldn't have dared to press those analogies, but I wanted to be on the #freeskip team.

#freeskip was exciting while it lasted. Then @colsonwhitehead moved on. Free Skip briefly became a blog where T-shirts and mugs were sold. And there was some self-congratulations among the Twitter crowds that revel in their brushfire discussion. As @dreamhampton put it in a tweet, "I love how Skipgate broke on Twitter first. And how we had the most sophisticated (and sarcastic) chatter about it." Then the #freeskip group dispersed.

Every day on Twitter is a list of blockbuster "trending topics" that come with a hashtag, and for much of the history of the platform they have been daily conversation starters like #itshardwhen and #iwannaslap. No matter what Twitter did or didn't do for the Arab Spring and other noble causes, it's also a wiki wit machine that specializes in one-liners. One Twitter hashtag, #moodkiller, inspired gags about the logistics of lingerie. Corey tweeted, "What a #moodkiller wen u tryna take off a girl bra n she got double d's n that last latch is holdin on 4 dear life." Don't find this stuff the height of humor? I kinda do. I always appreciate a guy's-eye perspective on the idiosyncrasies of bras. Finally, there's just something inspiring in the spectacle of an enormous crowd of writers working off a single idea to crack jokes and convey koans.

In 2010 Kanye West, a zealous user of Twitter, coined the phrase *hashtag rap* to describe analogies in rap lyrics that use a comic pause where the word *like* or *as* might be expected. West explained that this style conveys the rhythm of the pound sign in online hashtags. Consider these lines from the rapper Drake: "Two thumbs up [PAUSE] Ebert and Roeper" and "I could teach you how to speak my language [PAUSE] Rosetta Stone." Hashtags have become a big part of the rhythm of online communication on the social networks but even in email and texts. Leave it to West and Drake to formalize them as prosody.

Patrice Evans, a satirist and author of *Negropedia*, likes some of the hashtags on themes of race. "A good joke," he told me, "can walk through territory riddled with cultural landmines, engulfed in racial flamewars, and the laugh creates an insulated oasis. Jokes are like hurt lockers." At the same time, Evans isn't exactly blown away by the genius of most "blacktags," which he calls "Ghetto Comedy Voice." He is slightly bored by them, pointing instead to "white cultural fingerprints" in Internet prose. As examples, he cites "I. Can't." formulations and the bursts of ALL CAPS that he identifies with white bloggers and especially with users of Tumblr.

Baratunde Thurston, a *Daily Show* producer, comedian,

Snapchatter par excellence, and author of the best seller *How to Be Black*, once gave a showstopper talk at South by Southwest called "How to Be Black Online," in which he touched on the concept of black Twitter. ("What are my qualifications to talk about this?" he asked. "I have watched *The Wire*.") Citing marquee hashtags like #blackis and #blackaint, he told his Austin, Texas, audience, "You will find that Twitter is quite black in its culture."

In that speech Thurston addressed the much-discussed "digital divide" that suggests that race and class determine access to online resources. While black people may use home broadband less than whites or Hispanics, he explained, *mobile* Internet use among black people was for a long time considerably higher than for the other two groups. Mobile use, which includes mobile Twitter apps, lands black people in droves on Twitter, which many analysts myopically imagine as a place chiefly for white nerds.

NEVER NOT READING

As Jeff Bezos has observed, when you "consume" nonperishables like books, newspapers, electronics, and even the copy-dense megasite that is Amazon.com, you're reading. You're not eating, not consuming. In fact the signature pastime of the American consumer is now the mental act of processing digital, symbolic data: watching videos, graphics, maps, and images; listening to music and sound cues; and above all reading.

With media, books, texts, and emails on mobile devices people are never not reading. We read while we're socializing, working, shopping, relaxing, walking, commuting, urinating. From a nation that couldn't stop eating, we've become a nation that can't stop reading. As day follows night, our current form of overconsuming might be overreading. Hyperlexia. Reading texts while driving. Reading Facebook instead of sleeping. Buying multiple copies of books from Amazon, in print and digital form, as if to treat panic about future word famine, an imagined dystopia without text to read.

Over the past two decades screens have proliferated, filling our purses, pockets, and bedside tables. The living room is no longer configured around a single blazing digital fireplace, the television; instead it flashes with decentralized brushfires: ereaders, tablets, laptops, desktops, smartphones, televisions, refrigerator screens. As for the radios and bookshelves that were supposed to vanish with the digitalization of the American home, they've stubbornly remained.

No wonder shows called *Hoarders* and *Storage Wars* are part of what plays on our innumerable screens. The digital revolution that was supposed to sweep everything away—blast out the fat from our analog-and-paper arteries, as Nicholson Baker ruefully imagined it twenty years ago in his elegy for the card catalogue—has actually introduced more clog. It has distributed the clog between digital and physical without reducing it.

So it is utterly unsurprising to find that readers who embraced the ereader haven't yet burned all their traditional books and are, moreover, acquiring more. Lately I've found myself paying full price for digital *and* print editions of books I want to be sure always to have on hand. Not long ago Amazon announced a Kindle MatchBook program to allow customers to buy, for $3 or less, the ebook versions of books they'd already bought in codex form. Did you buy that hardcover of *The Secret*, with the chewy red seal on it, from Amazon in 2006? The same gibberish can be yours in eform for somewhere between nothing and three bucks. This is not a charity meant to incite literacy and save the publishing industry; it's a way to profit on the new and nervous hyperlexia, megaliteracy. This obsession is akin to the delusion (and actively promoted illusion) of food scarcity that has led, in part, to American overconsumption of food for forty years.

What does it mean to read—and screen—too much? At the biological and ethical level we're only now finding out. Hypotheses—alarmist, optimistic, and otherwise—abound. But at the economic level it means that we're buying copies of books we already have, suffusing our screens and bookshelves, along with our pantries, with hopeful redundancies, with stuff we don't need. Hoarding stuff, both binary and tactile. The American way.

3

IMAGES

MORE LIGHT

Early in 2014 a stimulating image worth at least a kilobyte of words circulated on social media. I admit I did my level best to hasten that circulation. When I pasted it on Facebook and Twitter, I was courting "Holy shit!"–style comments and ideally some context from Twitter followers of Facebook friends who might have a line on the image's component parts: China, pollution, high-def panoramic screens in public space.

The image showed a flat panel, ablaze in comet orange, mounted outdoors, like a drive-in movie screen. Around it hung a gray-brown sky, thick with aerosolized filth. A smattering of shadowy spectators and a clump of black-clad projector operators stood by. The tangerine radiance hit the heart as if it were solar—indeed the picture was an off-center sunrise over stratocumulus clouds—though it was the trompe l'oeil product of zillions of fierce light-emitting diodes, which evidently composed the screen. I say "evidently" because an image's power in social media derives from lab-made pseudo-language: the search-bait headlines, intros in Tweets, and traditional captions now optimized for search.

As likes and retweets piled up I both savored the collaborative frenzy of cultural transmission and began to nurse some doubts about the significance of the image. By chance it seems I was right to do so. A year later it was debunked: the sunset was in fact a single still in a public movie about something altogether else, shown in a public square.

Still the image itself recalled the arresting image of Funtwo eight years earlier. The pops of candy colors; the suggestions of Asian urban mysteries and semisinister piety about technology; the forcible muting of nature, only to bring back its simulacrum; and, of course, the haze. All of this made plain the brilliant uncanniness of digital imagery—and how deceptive it can be.

How did the Web, originally conceived as a system of hyper-*text*, turn so thoroughly visual?

It took some time, of course. Design, bad and good, had always wrapped our digital lives in the form of hardware and graphics. At first photographic images contained too much data to push through narrowband connections. But with broadband came Flickr, various photo sites, and YouTube. In the end it was mobile that did it, with Instagram and Snapchat, landing the whole history of *pictura* on, of all things, our telephones, and specifically our iPhones.

Just consider the iPhone 6 Plus, in pink gold, the first Apple phone to see its sales flatline. This is not a literate object. Though conceived in California, it has the debased European look of a duty-free luxury item or an object for sale in Dubai's brassily mirrored malls. It's also spacious, like a brushed-gold luxury mall in the desert: 5.5 inches. It's not a smartphone; it's a richphone.

Coursing through its hefty dimensions, moreover, is a royal road of zeros and ones: a broad, gracious data path best imagined as a Beirut Corniche of information. It's richer still that the chunk of raw value that looks like an actual slice of gold bullion opens only to the right biology: the fingerprint sensor perfected in the iPhone 5S. And what does the 6 Plus do, with all this fanfare and sentry system? Of that there is no doubt: the function of the phone resides in its brilliant screen and fast and ludicrously good camera. The iPhone 6—with engineers noisily insisting that the iPhone has become the most popular camera in the world—is a device for picture making.

Which means it's a device for making pictures in the hyperreal aesthetic of our time, characterized by vast chasms of range between sunshine and shadow. As a goal of photography, fidelity to the object has been left behind. Instead producers and consumers of iPhone photography seek a pointed visual high: the effect that Apple in its iPhone 6 promo video calls "more details in the highlights and shadows," the contrast, contrast, and more contrast known, in its most extreme form, as "high dynamic range."

It's an animistic look. The camera suffuses humble things with soul, spirit, and holy radiance. Household objects become incandescent in an iPhone photograph—to say nothing of our faces, irradiated as if haloed. The pupils of the Apple team can be seen to dilate as they make the sale: the iPhone 6's overbrimming pixels deliver 33 percent "more light" to images.

THE WORD

Only if you really strained could you hear, amid the latest whoops of new-iPhone hype, the soft sniffs of grief for another communication paradigm. As the iPhone 6 Plus drew the lust of Apple's overspenders, the once grand BlackBerry, that off-black fidget absorber that's still holstered by millions of small-time professionals, was wailing the last strangled notes of its swan song. I imagined I could hear in the mix the squeal and crash of an old fax signal, or maybe the squeamish-making "I Will Always Love You" ringtone from the 1990s.

There is an elegy in this. In the struggle for dominance between the so-called sister arts, poetry and painting—the struggle that is the sine qua non of Western culture—the demise of the BlackBerry points to the contemporary relegation of the written word to a supporting role.

This sidelining may be brief; certainly it is no cause for alarm. The exhausted written word must, like all phenomena, take breaks here and there—and not infrequently. A break for what the French call *langue* (language in symbols, not sounds) does not spell the start of a new Dark Ages. On the contrary, the mobile devices that have replaced the BlackBerry herald a glorious renaissance: of the picture, the photograph, the image.

But first let's mourn the BlackBerry, and its words—unlovely at times, but numberless. First adopted in 1999 by lawyers hunting for roving billable hours, then by government types, and ultimately by everyone lusting for hot fast communication, the BlackBerry brought the nation a new and nervous literacy: prayerful, chipmunk-like, addicted. Connectivity gluttons learned from the BlackBerry the ecstasy of onanistic thumb-wrestling with text.

For a time, it seemed, mobile technology was all about writing. And the love affair with digital writing started out possessive and turned openly erotic. Back in 2009, when President Obama brought his BlackBerry into the White House, he said, "They'd have to pry it away from me." This was thought to be the height of technophilia in the Oval Office. Love tilted into sex. Black-Berry fever soon ran so high so that no less than the Supreme Court heard a BlackBerry sexting case in 2010. In the spring of 2013 the FBI issued a memo officially asking feds to refrain from using their company-issued BlackBerrys to text-murmur "u r 2 sexy" to other feds.

Adults couldn't keep their hands or their passion off their BlackBerrys. The mobile-epistolary romances of the BlackBerry's heyday endangered careers and marriages. Mark Sanford.

Tiger Woods. Paula Broadwell. David Petraeus. Many of my friends, and yours. One by one, it seemed, public and private figures alike risked livelihood and household peace as they tested their Cyrano skills on their smartphones. They aimed to discover whether they could seduce with their thumbs and a grievously abbreviated character set.

Whatever else it was, see, the BlackBerry was a literate device, and it brought out writerly chops in unlikely Flauberts. For years the unspoken agenda of the device seemed to be the promotion of manic reading and writing. The BlackBerry foregrounded its twenty-six-letter alphabet and ready shifts to Portuguese and Hungarian accent marks, as well as, of course, arcane green-eyeshade squiggles: percent signs and pound sterling marks and *¢$#,àÜåß,àÇƒ©Àö.

That's why the BlackBerry worked so well for the so-called professional class, that grouchy clique of lawyers, bankers, doctors, professors, and journalists who no longer command the salaries or respect they feel entitled to. For years the advanced-degree set were the emailers, the memo composers, the readers of whole articles on nytimes.com. They adored the BlackBerry. Because even if all you longed to say, in the style of Tiger Woods, was "I really like doing that with you," if you were using a Black-Berry you got to say it in written language, the same way Proust and J. K. Rowling got to say their pieces.

But today the BlackBerry—and its wordy communiqués—is nearly extinct. Both TechCrunch and the bookmakers at Goldman Sachs insist that the iconic Canadian-made CrackBerry—once as enchanting as street coke—is going the way of crack itself. Research in Motion, the Ontario-based company that makes the BlackBerry, has for years been ceding market share, with unnerving Canadian politeness, to Google's Android and Apple's iPhone. Having commanded about half of the smartphone market in 2009, according to one estimate the company now claims less than 4 percent. In the age of radiant touch screens the ubiquitous, electrifying, and dangerous BlackBerry has become stale and unprofitable. The BlackBerry's braille-like keys—you could have written a novel on those high-functioning keys—now seem unhygienic. The keyboard is like a cluster of warts on an aged knee: nothing but traps for skin cells and bacteria and mortality.

In place of the morbid BlackBerry, Apple's smug, slick iPhone slipped in, a thief in the night. But this particular gadget shift, begun in 2005, has become more than a fickle market swap of old for new. It inscribes a profound psychic shift, from symbol to image. From word to picture. From verbal language to visual language, including emoticons, emoji, and photographs.

With its broad, rainbow-hued, semigloss face, the iPhone has no time for bespectacled phrase making. Like its brutal Buddhist taskmaster, Steve Jobs, the iPhone is dyslexic. Graphomania takes a holiday on Jobs's lickable, key-free device, which in place of incentives to write boasts extreme navigability, thrilling haptics, and an orgy of images images images.

I once asked Zac Moffatt, the smart digital director of Mitt Romney's vain bid for president, whether even Washington wonks were giving up BlackBerrys. "I hope so," he replied.

"BlackBerrys are highly efficient for people who are text-based. But people now communicate mostly in images, graphics, and video. For that BlackBerry puts you at a severe disadvantage."

People now communicate mostly in images, graphics, and video! Moffatt sounded so confident. "We are visual by nature," he explained, unprompted, identifying a cultural paradigm shift as if talking about biology.

Of all the explanations for the BlackBerry's demise—Apple is bent on world domination, diffident Research in Motion failed to adapt—Moffatt's account seals it for me most handily. The turn away from the BlackBerry and toward the iPhone is a reckoning with our essential nature and how we currently process, deploy, and enjoy symbolic communication. Look around: maps, video games, and news clips now surface on computers and phones where not long ago you would have expected text. Writing in the *Columbia Journalism Review*, Dave Marash designated online video history's first "universal language."

Witness the popularity of Flickr, Pinterest, Vine, Snapchat, and Instagram—all the services built for the exchange of visual digital artifacts. The increasing use of gifs, emoji, and images in text messages and on Slack is another form of visual discussion. Is this the truth of human nature coming out, or is this an odd contingency of mobile technology?

Possibly both. Sure enough, the iPhone lacks a physical keyboard. This is widely considered to have been a design choice by Apple; the one-button operation and broad, sleek touch screen has a loveliness that a device with fussy bumps and dirty crevices can't approximate. So the iPhone—and then the Android, which followed suit with a touch screen—is relatively hard to type on. But the iPhone has an extremely good camera that takes beautiful pictures.

Could it be that Jobs's obsession with sleek design—and perhaps his latent graphophobia, borne of his disability—led him to create a device that was accidentally word-negative? And then people who bought it for the music and the photos and the flash just gave up on text and slowly plugged themselves into new ways—visual ways, imagistic ways—of communicating?

INSTAGRAM

It wouldn't be the first time technology changed how humans swap symbols. When telephones were invented, people wrote fewer letters, and handwriting suffered. But phone voices became a point of pride: your phone voice was distinctive; your phone manner was distinctive.

When email first appeared, people stopped talking on the telephone so much; phone manners deteriorated to stutters, while online literacy soared. Now we are watching that literacy subside in favor of a *new* literacy. A new *visual* literacy. Ground zero for visual literacy is the Shiva-the-destroyer app, Instagram.

On April 9, 2012, when Facebook disbursed a cool billion

for Instagram, Jon Stewart couldn't believe it. "A billion dollars of money," he said, playing the analog crank. "For a thing that kind of destroys your pictures?"

Stewart's was a sound misunderstanding of Instagram, and he rendered it with more verve than the other Instagram misunderstanders who routinely take way too seriously the role of faux-vintage filters in the success of the photo-sharing app. To say that Instagram is just about blurred and tinted images is like saying that U.S. currency is about cotton-linen greenbacks inscribed with uncapped pyramids and all-seeing eyes. The value of Instagram does not inhere in the images, pretty or ugly as you may find them. It's in the deliriously complicated and heady circulation of those digital artifacts. The velocity. The trajectory. The way the ceaseless faster-than-light producing and transmitting, liking, tagging, commenting, and regramming can be leveraged for data mining and advertising in sets of encoded digital relations that make derivative securities look simple.

To the photographer, Instagram supplies camera, darkroom, processor, publisher, and a world of beholders in a single app. We snap the photos, then develop, crop, caption, display, and distribute our happenstance pixels by tapping the same device. Like tweets, Instagrams are indexed by both @persons and #subjectmatter. When you tag the heck out of an Instagram— laden it to groaning like a prayer wall at a Shinto shrine—you hasten and broaden its circulation by making it searchable.

And, yes, Instagram's filters can seduce. The coal-tar smears of Clarendon. Moon's extraterrestrial cratering effect. The ghostly central dodging of Amaro. Petal-pink Mayfair. The red-averse Lark. Hudson's winter sun. Sierra's misty reverie, Lo-Fi's rich and buttery glaze, Earlybird's Wyatt Earp sepia, Sutro's hauntedness, the Creamsicle kitsch of Toaster. Who can resist this ocular edge-taking-off? And the filters go on: Brannan's glamour and flash, Inkwell's perfect grayscale, Walden's incandescence. Finally, the amped-up, contrast-heavy Normal, which represents a mille-feuille of filters, but in a rich twist, now stands for unadulterated nature itself.

But don't let the artifice deceive you. Instagram is not an art project. Founded in 2010 by Kevin Systrom and Mike Krieger, two sumptuously capitalized Stanford grads, the app quickly became known for trademark clichés. Feet, skies, found-art latte foam, selfies. The tiltability and unobtrusiveness of camera phones, the dispensability of the flash, the simplicity of retakes and crops, and the one-touch lens flipper that incentivizes self-portraiture—these features evidently condition us to point our Insta-grameras in the same direction, over and over.

As in the best social media, the artifacts are not the innovation on Instagram. It's the system that's special. The name *Instagram*, it seems, does not so much play off *telegram* as *ideogram*. Instagram images have become units of speech, building blocks in a visual vocabulary that functions like a colonial patois, where old-school darkroom photography is the native tongue and digitization is the imperial language.

And like an empire at its height, Instagram is relentlessly making converts. Even people who don't yet use the app can recognize in its distinctive photos a new visual lingua franca. The images don't look filtered or like Instagrams. They look like reality.

The lessons of Instagram for social media and Wall Street are plain: the networks need apps to ride them. Clever social apps (Instagram, Vine, Hangouts) intimately tied to one network or another (Facebook, Twitter, Google) become a pretext for getting onboard those networks, for trailing one's data all over the Internet surveillance state, dragging one's eyes all over the advertising, and eventually clicking on this and that and spending good money.

But the effect of Instagram on individual users may be much more revolutionary. Now that superstylized images have become the answer to "How are you?" and "What are you doing?" we can avoid the ruts of linguistic expression in favor of a highly forgiving, playful, and compassionate style of *looking*. When we live only language—in tweets and status updates, in zingers, analysis, and debate—we come to imagine the world to be much uglier than it is. From Nancy Jo Sales's illuminating 2016 book *American Girls* I learned that many teenagers refer to their photo style as simply "aesthetic"—a meta-aesthetic meaning something like deliberately artsy. But Instagram, if you use it right, will stealthily persuade you that other humans, and nature, and food, and three-dimensional objects more generally are worth observing for the sheer joy of it. This little app has delivered a gorgeous reminder, one well worth at least $1 billion: Life is beautiful, and it goes by fast.

FLICKR

The Instagram story—and its role in the invention of a highly emotional digital visual language—has its origins in the story of Flickr, the wildly popular photo-sharing site founded by the Canadian company Ludicorp in 2004. Today some 3 billion photographs still circulate on Flickr, which was bought by Yahoo! a year after its founding.

Only by looking at Flickr can we see why the cartoonish distortions of Instagram filters became covetable in an art form—photography—that was once devoted to verisimilitude. It was on Flickr, after all, that the style of picture recognizable—and despisable—as "the Flickr photograph" originated. In 2008, while art school photographers were still shooting on film, embracing chiaroscuro, and resisting prettiness, Flickr's most distinctive offerings were the digital images that were believed to "pop" with the signature tulip colors that were then associated with Canon's digital cameras. (The tulipification of color was taken over by the iPhone starting in 2007.) While pretty and even cute, these Flickr images were also often surreal and prurient, evoking the unsettling paintings of de Chirico and Balthus, in which individual parts are beautiful and formally rendered, but something is not quite right overall. Flickr's creamy fantasy pictures, many of them labeled "erotic" (rather than sexy) portraits that have been forcibly manipulated with digital tricks, stand in contrast to the rawer and grainier 35mm photography that's still canonized by august institutions like the International Center of Photography.

Rebekka Guðleifsdóttir, one of Flickr's early stars, became the leading exponent of the site's style. Her rise to visibility also prefigured the rise of Instagram stars. An art student from Iceland, she learned to work Flickr before she became proficient with a camera. She discovered how to create the mini collections called "photostreams"—this before Instagram—and how to create images that would look good shrunk, in thumbnail form. She also learned how to flirt with the site's visitors in the comments. As perhaps is always the case with artists, Guðleifsdóttir's evolution as a photographer was bound up in the evolution of her modus operandi, a way of navigating the institutions and social systems that might gain her a following and a living.

Guðleifsdóttir's first photos were shot with an analog camera: snapshots of her school-age sons and a portrait or two of herself. Commenters loved the way Guðleifsdóttir looked— she's a weight-trained, protean-looking woman with movie-star eyes—but then Flickr members often deem 35mm photos "unfocused." ("A mixture of melancholy and curiosity," wrote a commenter on one image. "It's a shame about the focus.")

Guðleifsdóttir then shifted to a digital camera, first using a Canon Digital Ixus, then a Canon EOS Digital Rebel XT, one of the most popular cameras on Flickr in those days. (Discussions of cameras, lenses, and film pervade the site.) When she started uploading digital pictures, like her stony self-portrait *torso*, her photos started breaking Flickr records for numbers of views, and comments turned to catcalls. ("Gosh . . . huge breasts," someone noted.)

Just as Instagram photographers would do (after a fashion) years later, Guðleifsdóttir learned how to title and tag photos so that they might readily come up in searches, how to police copyright transgressions (as when some of her photos were sold illegally on eBay), and how to push contrasts by processing her pictures with Photoshop software. These skills might not have advanced her with New York galleries, but they made for a charmed ride on Flickr.

Her next step was to abandon realism. A few experiments in ten-second exposures led her to representations of specters and phantoms. Playing with shutter speed, she caught an image of liquor splashing out of a glass. She then started intensely manipulating and coloring her photos in postproduction, creating haunted interiors, doubled images, filtered landscapes, and contrived composites. Comments shot up; her page views hit the millions.

In June 2006, having followed Guðleifsdóttir's Flickr ascendancy, an advertising executive for Toyota came calling and assigned her a print campaign for the Prius in Iceland. She was to illustrate the car's hybrid quality, applying her wintry formalism and production mischief and producing many doubled self-portraits. A star was born, along with the legend of her rise to prominence on Flickr.

In the meantime another popular Flickr photographer who goes by the name Merkley was building up his own opus. His treatise extolling digital manipulation called "I'm Not a Photographer" derides mainstream art photographers who "show you shoes hanging on wires, pink boxes in the green weeds, little black girls with blue eyes and nuns sitting under billboards of naked men." On his Flickr profile he calls the classic film camera "The Robot Camera Machine" and proposes digital processing as the antidote to film's inhumanity.

Merkley's style is more R-rated and carnivalesque than Guðleifsdóttir's, but together the two Flickr stars mounted a case against vérité rawness in favor of posing, cropping, and special effects. Guðleifsdóttir and Merkley might have amounted to nothing in analog times, when elaborate deference to institutions, hard-won group shows, and expensive years spent in unnoticed toil were the only way to success. But just as certain ne'er-do-well writers have found themselves in blogging and failed filmmakers have taken to online video, these seemingly out-of-step artists have both created and mastered the Flickr photograph. Other photographers have added still more levels of processing—including the otherworldly contrasts achieved with high-dynamic-range photography—to the quintessential Flickr image, and it's becoming only more eye-popping and stylized.

And none of it looks like Diane Arbus or Henri Cartier-Bresson, the photographers critics still identify as the art form's greatest practitioners. On Flickr Cartier-Bresson is no Guðleifsdóttir. Maybe it's no surprise, then, that when a prankster posted as his own a Cartier-Bresson photo of a cyclist passing a spiral staircase, a mob of commenters shouted it down, crying for it to be deleted. "When everything is blurred you cannot convey the motion of the bicyclist," one commenter carped. "Why is the staircase so 'soft'? Camera shake?" asked another. "Gray, blurry, small, odd crop," someone concluded. That seemed to be the final word.

REAL TO SURREAL TO HYPERREAL TO FAKE REAL

Many of us grew up with the 35mm ideal of photographic fidelity to life's light and shadow. And we miss it, evidently: the moving midcentury Kodak portraits of late parents that pervade the graying realm that is Facebook express nostalgia both for absent friends and for a stable photographic aesthetic. The underlying idea was that a photo showed, above all, *what something looked like*. The camera was an eye for someone who couldn't be there in the Bahamas or at the baby's birthday party. Photographic prints show the faces of those who can't be with us now.

But those prints fade and distort. They were posed and produced to start with. And most important of all, all "realism" grounded in the confidence of art's "fidelity" to reality is a conceit of certain technologies. By the twenty-first century the confidence our parents and grandparents had in photography to tell the truth seemed itself antiquarian and quaint, so we *faded* our photographs to turn them into nonrealist representations of *realist* representations of reality. The result is a highly sentimental and unstable aesthetic, high on emotional power (mostly to promote positive emotions associated with cuteness, prettiness, loveliness, sweetness) and low on documentary power.

WHAT OUR CAMERAS ARE FOR

Nowhere is this shift in aesthetic more obvious than in amateur family photography. As any American with children knows, our children have at least one bright, clear reason for being: to furnish subjects for digital photographs that can be corrected, cropped, captioned, organized, categorized, albumized, broadcast, turned into screen savers, and brandished on online social networks.

I'm trying my hand at anthropology here: where farmers bred to produce field hands, industrial workers bred because they couldn't help it, and Kennedy-era couples bred to goose the GNP by buying sailor suits and skis, in the Internet age we form families so we can produce, distribute, and display digital photos of ourselves.

The marching orders come immediately, with the newborn photo, which must be emailed to friends before the baby has left the maternity ward. A conscientious father, chief executive of the budding business, must snap dozens of shots of the modestly wrapped newborn, generally with an Android or iPhone. He quickly examines the haul, with a view to how they'll play on Instagram or Facebook, scrutinizing pixels with the intensity of Anna Wintour. He selects a becoming one. The mother signs off from her hospital bed.

A parent may also edit the picture, correcting red eye or composition or even complexion problems, adding a filter or maybe animated confetti (depending on class affiliation). Enclosed in an email message, accompanied by a line or two of introduction, the portrait is broadcast like direct-mail advertising.

Thus a parent is minted. Good thing the drill starts early, as the signature act of Internet-era parenting repeats itself, again and again, in tighter and tighter cycles, throughout a childhood. It determines the rhythms of beach vacations and snow days. Eventually the business of family-image production and dissemination incorporates increasingly sophisticated and expensive cameras and photo-edit software and microblogging and distribution and organization systems (Instagram, Tumblr, Picasa, Picnik, MyFamily, Shutterfly, Snapfish). Before long the family has become a multimedia publisher, and—though it imagines itself a producer—a consumer of digital tools, gadgetry, and broadband.

For a parent this time-consuming vocation has twin payoffs: it wins you a break from your actual children while bringing you closer to their images. Pictures of kids, like idealized Victorian boys and girls, can be seen but not heard.

The child's life, reciprocally, becomes that of a model, and more. Every aspect of the family business becomes familiar to a child. Early on she learns that she can examine a photo on a viewfinder as soon as it's snapped; that she should monkey around rather than pose, as "film" is distinctly not at a premium; that a substantial share of her parents' mysterious clicking at their computers consists of organizing and reorganizing images of her. My own son's first word for laptop, when he saw a woman plugging away at one at Starbucks, was the word he used for himself: *baby*. What else could the woman be doing so intently at a screen but what he saw me do: paging through picture after picture of him?

The connection between parenthood and digital photography first dawned on me during Apple's video introduction of the iPad in 2010. In the videos a parade of Apple executives, clean-cut men with close-cropped hair, caress a glassy, oversize tile while proselytizing about it. "It's going to change the way we do the things we do every day," raved Phil Schiller, then an Apple vice president.

So which of digital culture's great offerings—which of the "things we do every day"—were most enhanced by the iPad? Are there shooter games, pornography, academic papers, live sports, message boards, chat, ecommerce, political blogs?

No. In the demos there were family photos. The iPad user, as we meet him, is a man alone, aswim in pictures of kids. Sure, he watches movies like *Star Wars: The Force Awakens*, reads business books, and plans and chronicles trips to Telluride, Colorado. But what he does most ostentatiously is organize and exhibit photographs of children: at birth, on the beach, in Paris in the rain, with conch shells to their ears.

One of Jobs's first boasts about the iPad screen? "People put their own photos on it." Later the iPad email client is demonstrated as it sends a baby picture. Scott Forstall, a senior vice president at Apple, doesn't hold back: "iPad is absolutely *the best* way to view and share your photos."

It's now half a decade later; personal technology is a delivery device for a lifestyle, a tacit prescription for how to live in the Internet's image-saturated symbolic order. Study the iPad or iPhone closely enough, and it seems to set a course for how we're now to use words and images for business and pleasure.

Maybe it shouldn't be surprising in vertiginous cultural times that the highest calling for the heaps of devices and services for the production and consumption of images—their cameras and virtual darkrooms functioning in excess of anything we rationally require—is to shore up our families and advertise them to the world and back to ourselves.

4

VIDEO

ME AT THE ZOO

Henry Adams said that to understand America, you must start with Chicago. It's central, non-European, and intimately connected to the nation's singularly abundant farmland and singularly ravenous financial markets that keep Europeans in awe.

Similarly, to understand video on the Internet you must start with YouTube. And, to be on the safe side, *end* with YouTube. YouTube, to which one hundred hours of video are now uploaded every minute, *is* online video.

The first video uploaded to YouTube appeared on the evening of Saturday, April 23, 2005. *Me at the Zoo* stars Jawed Karim, a trim, deadpan go-getter in an unzipped parka. Karim is one of

YouTube's founders. In the video he reports from an unnamed zoo—later identified as San Diego—looking into the camera shyly or maybe satirically. Children's voices can be heard and the bleat of a goat. Behind Karim elephants nose in hay. "Here we are in front of the, uh, elephants," he says. "They have really, really, really long"—a pause; is this double entendre?—"trunks." He turns to the elephants as if to confirm his observation. "And that's pretty much all there is to say." The video is nineteen seconds long.

Every citizen of the Internet should watch *Me at the Zoo*. This piece of ephemera is where the hypervisual Web started. The seconds-long video establishes some of the conventions that still define online video—which is to say, *all* video. *Me at the Zoo* limned the parameters and set in motion the most consequential way we now represent the world in art, in journalism, in everyday life, and in the intimacy of our mind's eye.

Internet video, for which the gargantuan YouTube is both Fort Knox and Library of Congress, is not just a powerful representational form, like breaking-news tweets, in which who-what-where-why words lay down a first draft of history. Internet video is our ranking realist form; it *is* history.

This is why Dave Marash in the *Columbia Journalism Review* went so far as to call online video our "universal language," as if the short videos that circulate on YouTube (and Vimeo, Daily Motion, etc.) were a fully transparent form of communication, without need of cultural context, curation, or translation. Ethan Zuckerman, the MIT media activist, took issue with this point, but in an important sense Marash had nailed it: more than any other form of moving or still picture, of language, of design, Internet video (short, digital, crisply hued, and infinitely shareable) registers in the contemporary mind as reality itself—the truth, history. The Record.

It's the credulity vested in the form itself that makes it so pleasurable to game. As I discovered over the years watching every manner of shock video, curio, and display of virtuosity, the first proper response to a great YouTube video is "Holy shit!" And the second is "This is a fake!"

The video-sharing service YouTube was the inspiration of Karim, along with cofounders Steve Chen and Chad Hurley. (The Pete Best of online video, Karim parted company with the others and was left out of YouTube's runaway success.) The three men had been colleagues at PayPal, the online payment service. Having cannily weathered the dot-com crash of 2000, PayPal and its parent company, eBay, were considered paragons of spectacular and durable success on the Web.

The trio had learned PayPal's lessons well. With YouTube they stuck with PayPal's primitive and even homely nondesign. Hurley's chief graphic innovation was to smack the "play" arrow right in the center of the video window, a design move that weakened a viewer's resistance to watching by making a video seem all but already under way. The arrow enacted a "conversion," in the idiom of ecommerce: it turned a lurker or surfer or browser into a participant, a customer, and eventually an attention span for prerolled ads.

YouTube's founders also employed stealth nonmarketing. No premiere hoopla marked the debut of *Me at the Zoo*. YouTube decidedly wasn't putting on a show, busting into the rarefied and risky entertainment business. Instead it was creating a commodities exchange, a place to set standards and establish trade rules for swapping video, as was done in the original American mercantile exchanges of the mid-nineteenth century, for butter, cheese, eggs, soy, and alfalfa. With its famously polyglot and boisterous comment section, YouTube is still an open outcry trading pit, meant to enable the high-velocity, high-volume trade of digital artifacts.

That's right: exchange. YouTube was never in the entertainment, education, or publishing business. This alone was a silent breakthrough. In 2005, at the advent of Web 2.0, when dialup receded, bandwidth ballooned, and social media and online video came to define the consumer Internet, Hollywood producers and New York publishers, and especially the many graduates of liberal arts colleges who aspired to their cultural status, dreamed that the Web would be a new place for trained professionals to entertain and inform respectful, paying viewers and readers. The coastal cultural establishments in the United States weren't thinking about exchange. That was a job for Wall Street and Chicago. For the merchant class. For their part, the culture folk who scorned law and medicine every bit as much as finance and sales—the prosperous artist-professionals with their BAs in Brown's semiotics or Harvard's "Hist and Lit"—intended to use the Web to do what their crowd had been doing for one hundred years and more: make books, newspapers, magazines, movies, radio, music, and television.

By contrast, the twentysomething YouTube founders held no brief for Condé Nast or Paramount Pictures, much less for the *Paris Review* or the sacred cows of American filmcraft like, say, Robert Altman or Woody Allen. Two of YouTube's founders were not even raised among these American cultural totems. Born in Germany, Karim is of Bangladeshi German descent. Chen, who is Taiwanese American, was raised largely in Taiwan; his family moved to the American Midwest when he was fifteen. All three of YouTube's founders had studied computer science, and not one was poring over the yellowing playbooks of twentieth-century media empires. Karim, Chen, and Hurley didn't fantasize that a digital offshoot of legacy companies could serve the online world. Instead they saw the Web as an entirely uncultivated land, still in dire need, first and foremost, of posts at which to gather and trade.

As teenagers and collegians in the 1990s, the YouTube founders were also mindful of the bright encampments that constituted Web 1.0: Amazon (launched in 1994), eBay (1995), Craigslist.com (1996), and Match.com (1996). For a time these sites could pass, comfortably, as Web analogues of offline businesses (a bookstore, an auction house, personal ads); they thrived on hypomanic levels of participation, community, and sociability. They required a colonial ethos of collaboration and gratis work, or no one would get fed.

Michael Hirschorn's Facebook axiom that I cited

earlier—"You have to give information to get information"—admired the dues asked by the web for those who showed up without pitching in. Those dues were there from the start, as those shy of computers were lured to the Web with, in part, the threat that they'd *pay* by missing out on chances (for romance, employment) and deals (on books, used furniture).

The first social free-for-alls like YouTube were closer, then, to the Chicago futures market, with barrel-chested, red-faced men pricing soy in those squalling pits, than to Hollywood studios with their unions, auteurs, and steep barriers to entry. Before YouTube the new territory of the Web had only elemental needs. It required zones for trade, social life, and peer-to-peer showmanship in which digital artifacts, like eBay's stars or Match's winks, would function like PayPal's currency.

The coin of the YouTube realm would be video. But not because anyone at YouTube loved film or even twentieth-century television, which inspired the company's retro logo. Video just happened to be on hand. The same way the Warner brothers of London, Ontario, started with a projector in 1902 and found themselves needing to fill the patch of projected light with something worth watching (*Life of an American Fireman*), so the YouTube brothers found themselves with a display and distribution service and needed something to display and distribute (*Me at the Zoo*).

And video was around: by 2005 young people and new parents had been emailing videos (parties, stunts, babies) made on Canon and Nikon compact DV cameras to family and friends, who suffered dial-up and buffering headaches to watch. It was exactly the emailing and the headaches that the YouTube trio hoped their service would replace. They aimed to pick up the pace too. As they knew from PayPal, it is *velocity* that determines a sophisticated global market in the age of derivatives and currency exchanges, more even than static measures of supply and demand.

This aspiration was nothing less than to build a new economy with a new currency. The chutzpah of YouTube's ambition jibed with something told to me in Hollywood that same year. Jon Katzman, a onetime executive at NBC and Warner Bros., who'd helped develop *Saved by the Bell* and *The Fresh Prince of Bel-Air* in the 1990s, had invited me to a good-natured pyramid-scheme "festival" for mobile video in Beverly Hills. (In the mid-2000s everyone interested in Internet video, an unknown proposition, was perforce bluffing; Katzman admitted he had invited broadly to his first Third Screen Festival, using one name to draw another, without ever being sure what the event would offer.)

In a revelatory discussion in a windowless office Katzman told me he had realized his own path was diverging from that of his father, a movie producer who'd considered his life's work "to tell great stories." His father was spinning fantasies for viewers in the twentieth century—of glamour, more often than not, or romance, adventure, valor, delight. As I listened I recognized this traditional prime directive as the same one that drove the publishing magnates I encountered in New York, people who prided themselves on first bringing *Lolita* into print, or creating *Allure* magazine, or assigning Janet Malcolm her first magazine piece.

But Katzman Jr. had himself lost interest in the "telling great stories" pose. This sounded like heresy to me—I wasn't ready yet to swap the ideal of the story for the ideal of the system—but I paid attention. Katzman had recently been up north, to San Mateo, California, home of YouTube, and there—with distance from L.A.—he had gotten religion. This is a story I would hear many more times in the coming years. The great-man producers of our own time, Katzman explained, were no longer the raconteurs of stage and screen, permitting viewers the fantasy that they were John Wayne or Cate Blanchett. Rather the new great-man producers were creating platforms that would permit others the fantasy of *auteurism*. A memory flashed before me: my brother and I as kids, arranging figures in front of our dad's stop-motion movie camera, imagining we were not Luke Skywalker but George Lucas.

These platforms would allow *other people* to tell stories, great or otherwise. If you owned YouTube, the storytellers were the audience, the consumers. The storyteller was no longer controlling things. The great-man storyteller was, in fact, the new chump, the new sucker, *the one who would pay*. Telling stories was no longer producing; it was consuming—bandwidth, technology, platform space, code. And the storyteller was the consumer who would eat what you fed him and pay for more. If the storyteller used to be feeding consumers stories, now it was the man with the *system* who was feeding storytellers bandwidth.

You, the platformer, Katzman explained, as I warmed to his enthusiasm, had the master story, the metastory, the wordless story that undergirded the illusion of personal storytelling for the consumers: you owned the *system*. That's what Katzman said he wanted to do, or anyway, for now, what he thought was most worth doing.

This was prescient, and it stuck firmly in my mind. After all, if Katzman was right, this inversion of production and consumption, or the blending of the two, would completely change what would count as cultural work. Katzman said this years before the *Huffington Post* appeared and demonstrated that writers not only would write for no payment but that they'd themselves pay—in labor and expenses—for space on a digital platform. I now imagine the infrastructure of vast HuffPo and far vaster YouTube as the award-winning showpiece on the stage, and the rest of us—contributors of videos and words—as merely audiences spellbound to the code. We pay to behold that code and bask in its reflected glory every bit as much as the first movie audiences paid to behold the Warners' machine and its mysterious beam of variegated and changeable light.

LONG TRUNKS

Back to the first YouTube video and its fixation on elephant trunks. When this technique of redundancy was used in the films of Jean-Luc Godard, it was considered the height of New Wave sophistication, a commentary on the way movies pile on information. They show, they narrate, and they describe. The elephants are unmistakable to viewers, and yet Karim identifies them. Then he names the iconic shape right in front of us—"long trunks"—lest anyone miss that long trunks equal elephants equal long trunks.

This founding clip makes and repeats a larger point, too, with every pixel: video, trivial or important, could suddenly be published, broadcast, and shared, quickly and at no cost. *Me at the Zoo* also set a style standard for the classic YouTube video: visually surprising, narratively opaque, forthrightly poetic.

After the zoo, the deluge. A decade later, YouTube, now deeply integrated into Google, takes pride in its eye-popping metrics. Here's one: If all three major American TV networks (NBC, CBS, and ABC) had been broadcasting for twenty-four hours a day, seven days a week, for sixty years, they wouldn't have created the amount of content uploaded to YouTube in two weeks.

So the video floods in and YouTube is an exceptionally well-oiled machine. But the content continues to be enigmatic. While YouTubers in the early days often produced recognizable short-form genres that existed on TV—music videos (*Numa Numa*) and sketch comedy (*MySpace: The Movie*)—uploaders since have drifted from known forms, contributing entries now known only as "YouTube videos." That's because it's not clear what these videos would have been called before the advent of the site.

Homemade oddities that attract attention in YouTube's second decade include numberless pet videos, heartland vignettes, security-cam film of debatable authenticity, stunts, accidents, and DIY animation. There are also clips from actual television, with and without the permission of the channels, and copyrighted, professionally produced music videos uploaded by outfits like Sony and Universal. Somewhere between the weirdness of the homemade videos and the possible illicitness of the professional stuff (which might violate copyright), YouTube hit gold.

This is another component of Internet civilization: consuming art here feels like truancy, like something shady, something you shouldn't be doing. And this is the best effect art can have.

It's elementary semiotics. Measures taken to keep images from being seen—veiling, blurring, coding, scrambling—often become as exciting as the images themselves. Sometimes more. Photoshopped pinup girls lose allure when staring at them on billboards comes to seem mandatory. Sooner or later anyone seeking intellectual or physical arousal wants to see something she's not supposed to see. YouTube videos derive huge appeal by seeming louche. A drunken celebrity, a failed stunt, a politician's outburst on that crazy and cluttered site: Should I really be watching this? If you feel a little remiss while watching a short moving picture online, chances are good you're watching a YouTube video—and it's having the desired effect.

People contribute to YouTube because they want to tell stories, be heard, be seen, be known and maybe famous. But why do we watch short, helter-skelter videos that for years were so arbitrary and unclassifiable as to be mostly maddening? Because haikus and weird miniature ivory carvings and a second's inter-action on the street can be exciting. Flashes of rage are excit-ing; displays of magic or talent or turns of mood are exciting; instantaneous emotional contact is exciting; and the suggestion that anyone with a cheap camera, stuck and lost and jabbing an-hedonically at keys all day (like all of us), can appear to the world through those screens is exciting.

At the same time, just as YouTube was revving up, tradi-tional sources of entertainment and art—the music produced by the "music business," blockbusters made on the Hollywood model—were becoming nearly moribund. The opposite of ex-citing. In YouTube's early days the biggest growth sector of the music business was in how cell phone companies were forming cartels with record labels to overcharge for hip-hop ringtones so they could cobrand with content-aggregator initiatives to cre-ate entertainment verticals that allow for parallel advertising of other ringtones in the name of being "interactive" or doing con-sumer-to-consumer programming. Was it any surprise, given the despair latent in the legacy cultural businesses, that people turned to YouTube for signs of actual life?

SOLD

A poker-faced Chad Hurley, then twenty-eight, coiffed like Liszt, held court with some *Times* writers in the fall of 2005, just before—without warning—he and Steve Chen sold YouTube to

Google for $1.65 billion. Young Hurley, with the sangfroid of a surfer or a seasoned Wall Street shark, didn't mention looking for a buyer; he didn't mention numbers or stock. Instead he told a room of media writers that YouTube didn't exist to produce or even broadcast video; rather it enabled the exchange of video. He kept using an odd nonbusiness word: *share*. In this way, as Hurley put it, YouTube was a shadow social network, like Friendster. (Facebook, which was founded the year before as a Harvard-only social network, was not mentioned.)

I asked Hurley about pornography. I had been writing about Internet video for the *Times* for about six months, but I didn't yet consider YouTube video's base camp. That was because many other comedy and shock sites, like Ogrish.com (which began showing stills of World Trade Center jumpers right after the 2001 attacks), Gorillamask.net (founded in 2005 and named for a weird sex act), and Ebaumsworld (teen boy comedy site with a *Mad Magazine* aesthetic), showed plenty of video. All these sites still seemed like contenders for the role of master video depot. At the same time the shock-and-comedy sites seemed to be drifting steadily to sexy vids, their homepages increasingly dominated by naked girls. I assumed YouTube would go the way of those other sites, homogenize entirely with nudie stuff, and fulfill the prophecy made in song in 2003's *Avenue Q*: "The Internet is for porn."

Hurley explained that YouTube had a secret process for keeping the site clean. We asked, awkwardly, if an algorithm yet existed for shooting down flesh-colored pixels—reporters were just beginning to try out words like *algorithm*—and Hurley smiled, faintly. Years later, in 2009, after a breach of YouTube's decency standards by some 4chan hackers who uploaded pornography passing as music videos by the Jonas Brothers, I learned that it's users who flag offending videos, which are then screened by a reviewer and struck down when called for. That's right, no "algorithms": Google's video empire just waits for citizen censors to sound alarms and then have porn spotters press "play" on each and every video. *Yep, that's porn, all right. Nope, that's a breast exam.* The 1964 epigram of Potter Stewart, the Supreme Court justice, just won't quit. Obscenity: we *still* can't define it, and we *still* know it only when we see it.

Scott Rubin at YouTube explained to me that the site does augment its community-flagging system with proprietary screening technology. This technology vets videos that have been flagged and helps the human reviewers prioritize which ones to watch. Rubin wouldn't tell me exactly how the technology works, so I ventured some guesses. Does YouTube's software read the descriptions of videos that are provided by uploaders (flagging words like *hottt* or *sexytime*)? Does it trace the origins of videos to shady IP addresses? Or does it do something harder, like zero in on pinkish pixels?

"I could have a video of my family at the beach where there's a lot of running around and a lot of flesh," Rubin said, "and of course it's not pornography. It's probably just boring pictures of my kids running around." In short, he added, "technology can never do the work of our reviewers, because video is always contextual. We try to determine the intent of the uploader. Someone uploaded a video"—staged for a public-service announcement—"of a teen giving birth on a playground. It stayed up. Reviewers watched and concluded that the playground birth had educational, documentary, or scientific value. Technology could not do that."

Regulation, it seems, requires the contribution of a human eye and a value system. This should be no surprise. Material contraband like drugs and guns have chemical and physical hallmarks; they can be sought by those properties. And you can, of course, ban strings of letters from any textual material that allows for find-and-delete procedures. But while images can be indecent and even illegal, the component parts of those images—the pixels, the light and shadow—are, taken in isolation, pretty meaningless.

No doubt this is why "tags"—words connected to an image—have become so integral to sites like Flickr and YouTube. They represent an effort to translate images into language. A beloved YouTube clip of Norman Mailer in 1971, for example, is tagged "Norman," "Mailer," "feminism," "cult," and "bloody," along with other words. Those familiar with the clip will know why. Many sites allow all users to tag images, drawing on the wisdom of the crowd to create a "cloud" of associations.

That day at the *Times* Hurley made it clear that porn was not going to enliven YouTube anytime soon. This was a surprise. Though Jobs's anxiety about vulgarity was well known, showing up in some semicensorship on iTunes, it seemed unlikely that young guns like the YouTube team would pander in this way. What's more, it had long been accepted as axiomatic that technology and pornography—from the printing press to photography, magazines, film, and videotape—always evolve in tandem. "Sometimes the erotic has been a force driving technological innovation," John Tierney wrote in the *Times* in 1994. "Virtually always, from Stone Age sculpture to computer bulletin boards, it has been one of the first uses for a new medium."

Hurley didn't seem to care that porn was where the money was, or the innovation. Instead he seemed to see it as I came to: as a pixilated kudzu, a plant charming enough in itself but rapacious in an ecosystem and capable of choking off other kinds of growth. The predator plant I described earlier.

YouTube was determined to cultivate *its* ecosystem in such a way as to remain, above all, diverse, brimming with video organisms of every known genre and microgenre, and millions more we'd never seen before. Before YouTube who'd ever heard of a mashup, a "fail" video, a "haul" video, nip slips, lolcats, Auto-Tuned news? Just as making YouTube a site for exchange rather than broadcast was ingenious, banishing porn clinched its status as the video-world frontrunner. Unlike movies or TV, online video, YouTube taught us, was meant to be shared and meant to be infinitely heterogeneous.

At the same time, in the mid-2000s YouTube was suspected of other forms of censorship. It's a sign of the times that, in 2006, when right-wing firecracker Michelle Malkin complained that YouTube was just more jihadi-loving mainstream media in sharing-and-caring hippie clothing, the charge—made by Malkin on YouTube—was reported by both Tom Zeller Jr. and me in the *Times*. A decade later who could keep track of the ideological videos uploaded to YouTube, tendentious or incisive or both, much less report on them?

But what was the content and form of these video artifacts that were beginning to whip around the Web? Were they news clips, *Saturday Night Live*–style sketches, or gotcha videos of police asleep at their posts? No, it was amateur silliness, seemingly. Placeholders. Messages that existed, like iodine in veins, only to show the networks in action.

At the same time, amateurism and accidents and gotchas became an aesthetic of their own. Music videos, a form once considered nearly obsolete, made a comeback precisely because they express a very contemporary, entirely YouTube style and even ideology. In short, the best music videos (for YouTube, as opposed to MTV) are garish, clumsy, erratic, dirty, and densely allusive; they seem to come from the margins, back alleys, and black markets of commercial culture.

Take Beyoncé's masterful *Pretty Hurts*, which beckons with YouTube's irresistible come-on: watching this should be forbidden. The video, in which Beyoncé competes as Miss Third Ward in a regional beauty pageant, is pervaded by forced vomiting, modern corseting, female mustache-waxing, and other nauseating "Rape of the Lock" boudoir rituals. A leering Harvey Keitel makes an appearance as emcee. The video's impression that it shows the clinical and slightly sickening underside of commercial culture—the not-ready-for-TV and not-meant-to-be-seen side—is gold.

It didn't surprise me that *Pretty Hurts*, with its gritty and even foul undercurrent, was the work of Malina Matsoukas, the ingenious director responsible for *Why Don't You Love Me?*, an earlier Beyoncé video that encourages interest with a similar conceit of viewer trespass. Anyone who still thinks video commenters are Neanderthal masturbators should go back and see how scholarly things got early on with that one. "This is a homage to Bettie Page," one commenter declared, pointing readers to a vintage Page video and noting that "the intro music comes from a 50s film." Another wrote, "Modern fashion played vintage!" And another, "The vintage color treatment is so spot on!"

With that video, Matsoukas told me, she and Beyoncé wanted to create a video inspired by Bettie Page movies "without telling anyone—not her label, not management, not anyone." They chose the right medium. Online video always seems as if it's going behind the backs of managers and labels; the story of the video's creation complements its scrappy aesthetic. With its grubby yet painted surface and its boozy, ashy mise-en-scène, which positions an unstable housewife and her toxic cleaning products at center stage, *Why Don't You Love Me?* trails the scent of Shalimar and Tilex just as surely as *Pretty Hurts* reeks of spray tan and Minx nail glue. The density of these videos implies that they've got secrets. Justin Wolfe, of the blog Firmuhment, argues that music videos like Beyoncé's jam themselves with clues expressly to inspire the forensic madness in which the Internet specializes. These videos, Wolfe writes, are "complex and multivalent and open enough to stand up to the sort of lengthy exegesis that people do of TV shows, the kind of amateur critical study that you can build communities around and write a million blog posts and comments about." (For further study, Wolfe cites *Run This Town*, the video with Rihanna and Jay-Z, which contained suggestions of Freemasonry that burned up the brains of online close readers for a time.)

But Joycean density isn't all it takes to make a music video tantalizing on the Internet. It also needs Joycean sex. Many music videos now appear in two forms online: censored and not. The doubling effect suggests that someone like Beyoncé or Lady Gaga or Christina Aguilera keeps it clean for TV (and when she's being "real" on Twitter and in tabloids), while after hours, if you can find her on the Internet, she's infinitely dirty. Matsoukas played with the clean-dirty axis at the level of the film stock for *Why Don't You Love Me?* "We shot on Super 8-millimeter and 16-millimeter (for safety)," she told me in an email message, "and later on went and 'dirtied' up the 16-millimeter to match the Super 8-millimeter by adding grain and dirt and defocusing the sharpness of the images."

The Gaga-Beyoncé *Telephone* video used grainy images too, especially faux security-cam film in a stagy women's prison, which really made the images seem like contraband. It's pretty thrilling that music videos can get that dirty, exciting effect again; it might even reawaken us to the power and pleasure of film.

With the "You" in YouTube—the DIY uploader—whole genres are being invented. Jean Cocteau once said, "Film will only become an art when its materials are as inexpensive as pencil and paper." Without anyone noticing, on YouTube film had become an art.

It's still surprising today how little the homemade videos on YouTube resemble the pro goods. Sure, there are parodies of mainstream clips here and there, but mostly the amateurs are off on their own, hatching new genres.

Consider "haul" videos, in which people show off the stuff they recently bought, or the evergreen "fail" videos, which show all manner of human endeavor gone wrong. Individual haul and fail videos often attract 100,000 views or more, and no one had even imagined such genres until recently. Yet no one at any production company seems to be struggling to serve the haul-fail audiences (or combine them?). And the haul people and fail people evidently don't feel underserved; they are helping themselves and creating what can only be called an art scene, all around the many, many videos of their genre on YouTube.

In serving these niche audiences with their microgenres, YouTube has solidified its slot as a home for the vernacular avant-garde. For years I have believed this, and for years people have warned (or promised) me that any day now the heterogeneous site would be steamrolled by commercial forces that would wipe out the indigenous flora and fauna. But not only has the weird, small stuff hung around; it also continues to be found by its audience. YouTube may not be making money as efficiently as

Google once hoped it would, but it's still incubating novel forms of creative expression and cultivating new audiences.

This hit me when a YouTube video called *Manhattan Bridge Piers* made the rounds on TV and blogs. A beautifully shot, silent, time-lapse documentary, it shows the Manhattan Bridge dramatically wobbling when the subway crosses it, as if on the verge of collapse. Commenters came at the video from a range of angles. "Amazing video!" one wrote. "Fat New Yorkers!" snapped another. "It's designed to be flexible. No structure of that size can be made completely rigid," wrote a third, whose assessment was borne out by professional engineers who commented elsewhere.

In fact, though *Manhattan Bridge Piers* was played on TV for shock value, the video, which was uploaded by a YouTube member named kvertrees, whose work includes other silent films, of mushrooms and swallows, works better as a tribute to suspension technology. And it works best as an art film. It recalls nothing so much as *Manhatta*, another silent documentary, from 1921, by the painter Charles Sheeler and the photographer Paul Strand. With title cards that quote Walt Whitman, *Manhatta* turns the urban landscape into a series of painterly, often abstract vistas. Art museums now show it alongside the still photography and painting of Modernist masters. As in *Manhattan Bridge Piers*, the city seems ominous in *Manhatta*, if only because it so completely overshadows the humans in it.

No wonder some viewers envisioned literal collapse after seeing kvertrees's film. In *Manhatta*, which emphasizes how the city dwarfs and threatens its inhabitants, a crowd of commuters disembarks from the Staten Island Ferry; they're jammed so tightly that they seem in danger of suffocating or stampeding. Similarly there's virtually no way to see the city in *Manhattan Bridge Piers* except as one secured by fragile pacts. In both films the filmmakers align their technology with soaring modern structures and seek vantages from which to underscore that identification. But the humans in the two are mostly out of sight, encased in monstrously beautiful architecture that can be truly beheld—in all its smoking, soaring, swaying, flashing sublimity—only by gods or cameras.

None of kvertrees's other videos, each powerful in its own way, has received the attention that *Manhattan Bridge Piers* did, largely because news outlets couldn't peg a discussion of infrastructure disintegration to them. But the others, the mushrooms and the swallows, are worth watching too. As casual users puzzle over YouTube—its mayhem and trivia, its commercial and political uses—hard-core users quietly, steadily figured it out. And it took only ten years, from *Me at the Zoo* to the hundredth hour of *this very minute*'s haul of videos uploaded to YouTube. For better or worse, YouTube is a place for art.

ULTRAHD3D: THE IMMERSION

Art in the digital age might be defined chiefly by UX. When UX is a term of art, it's an engineer's word, prototyped with a blueprint-scaffolding system called wireframing. More broadly applied, UX refers to user responsiveness. There are various ways to think of this, and one useful way is somatically—by consulting the body.

This is a serious question: Do you like your art in front of you, inside your body, or all around you?

People who are drawn to plays, movies, and TV sports generally prefer to keep their diversions at some distance. These are traditional "viewers." We sit in relative shade or even theater darkness while our entertainment is brightly lit or backlit. People who like food, perfume, and music in headphones like entertainment in their mouth, nose, and ears. They are cultural "consumers" and generally take their pleasures in low restaurant lighting. People who like architecture, video games, music in speakers, and, most recently, 3D media seek to be surrounded and included in the action. They like a diverse, changing light scheme, like those in cathedrals, theme parks, and dance clubs. In marketing lingo these people are "experiencers."

The case for 3D is often framed as a case for realism; 3D seems truer to life than flat 2D. On that logic 3D is part of an aesthetic evolution that has included the shift from medieval fixed-point perspective to illusionism in Renaissance painting. This makes sense. Two-dimensional art always needs devices that allow it to more convincingly suggest a third dimension. Why should movies and TV be any different? Even though 3D technology so far is largely associated with kids' movies, 1950s kitsch, and novelty for its own sake (*Up*, anyone?), it is not intrinsically childlike.

But there's also a psychological component to the case for 3D. "Depth" is not just geometry, another line on the cube. We have strong emotional attachments to depth. Art without any effect of depth can seem heartless or superficial. The absence of shadows and other visual depth cues from soap opera and game show sets and the use of lighting fills to erase wrinkles build a monochrome palette that suggests that daytime TV lacks moral heft and consequence. By contrast, shows like *House of Cards* and *Breaking Bad* are loaded with depth cues, including marked contrasts of light and shadow. No wonder these shows strike us as profound—*are* profound.

For a sense of how deep 3D might one day be, look to recent low-budget 3D film and TV projects. These are not thrill-a-minute blockbusters for the roller-coaster set. They're documentaries and independent fiction films. As Jed Weintrob, now the head of production at Condé Nast Digital, told me by email, "Most people assume that 3D works best for 'big' event films, concerts, and IMAX-style documentaries, but I have actually found that some of the most gorgeous 3D imagery has been in 'small' dramatic scenes and intimate moments in the real world."

Editing broadcast-quality high-def 3D is an intricate process of making two digital "reels"—painstakingly shot by the company's proprietary cameras—converge, almost. Achieving this near-convergence ensures that, when seen through glasses on a new 3D TV, the image appears beautiful and not nauseating. (Though new glasses-free small screens like receive relatively high marks from Weintrob, he points out that "displaying

3D will always involve sending each eye a different image," which presents technological challenges for screen design; thus "a glasses-free big screen is still a ways away.")

Watching one of Weintrob's films I was sufficiently dazzled. The film documents, among other things, a voyage up the Hudson River of the *Clearwater*, a boat built by Pete Seeger, who died in 2014, as a replica of a nineteenth-century sloop. Without the glasses the image looks like one you'd see drunk. With the glasses the ship comes into brilliant relief, but the sparkling river is the real showpiece. The Hudson looks as it must look to sailors on calm days: a highway so supportive and broad and horizontal you'd think you could walk on it. In 3D a river does not look like an extension of the sky, the way it often does in movies and watercolors; it looks like an extension of the earth, with which it shares a plane. A revelation.

I find movies like *Avatar* and *Gravity* enlightening in 3D, partly for the filmmaking and partly for the vertigo. It turns out I'm among those who have problems seeing 3D movies; we're 5 to 10 percent of the general population, according to Marc Lambooij of the Eindhoven University of Technology in the Netherlands. My 3D problem registered as mild motion sickness, which didn't let up till the closing credits.

WASTING OUR TIME

Over the decade I wrote several times a week about television for the *New York Times* there was never a time—and there had never been a time—that people watched TV without thinking they were *wasting their time*. Many people I knew in New York told me they didn't watch TV. They hadn't watched as children. They wouldn't let their kids watch TV.

It's true that television has historically defined itself against work. The sedentary television rules the living room in view of a fat sofa, those great monuments of domestic leisure. The evening news marks the end of the workday; prime time exists as an alternative to plays and movies; and those lucky enough to be sick or jobless get to watch talk shows and soaps. Watching television is so much the opposite of work, in fact, that it's hardly even a purposeful act: if you spend Saturday and Sunday on a Netflix binge, you can credibly say you spent the weekend doing nothing at all. Wasting your time.

When I worked in the Arts section at the *New York Times* I worked with critics who wrote about film, dance, books, opera, theater, and art. I decided I had it easy, except my subject, television, was considered, in a steady stream of inconclusive but also frightening reports that the paper regularly covered, a public health hazard. Poison. Dangerous. Likened often to cigarettes and junk food. Read any book you like, no matter how silly; watch any movie, no matter how violent; see any play, no matter how boring or bawdy—but if you watch even fifteen minutes of TV it's doing damage to your body, mind, and soul.

Television, after all, is a *vast wasteland*. Newton Minow, the FCC chair, had called it that fully fifty years ago. He had issued a curse on the whole medium when he said it, but also a challenge. At least that's how I came to see it: TV, and then digital video, has to operate under the sign of disreputability. And I decided that was the best thing about it. I had trained, after all, to be a literary critic. I knew that in every period there is an art form—satirical poetry, *comeddia*, Hollywood movies— that is disreputable. And that people can't stop consuming. Like Shakespeare. Like Whitman. Like Nabokov. I knew that a cloud of guilt is good for art.

As I told Mr. Minow when I met him on a panel at Northwestern, the best thing anyone has ever done for television was to curse it, right then, right at the start. He put TV under a moral cloud, and its efforts to burn off the cloud—but also to wickedly enshroud itself, like Huckleberry Finn saying, "All right, then, I'll go to hell"—is what has made TV the great medium of our time.

It's not just that the past fifty years have witnessed some great moments in television; it's that the past fifty years have seen the greatness of television in its entirety. It has been what the nineteenth century was for novels and the Romantic period was for poetry. *Homeland. The Good Wife. Orange Is the New Black. Parks and Recreation. UnReal. Transparent. The Affair. Breaking Bad. Mad Men. 30 Rock. Bloodline. Friday Night Lights. The Hills. Deadwood. Entourage. Rescue Me. The Comeback. House. Weeds. The Office. The Colbert Report. Nip/Tuck. The Wire. Arrested Development. Six Feet Under. The Sopranos. Will and Grace. Sex and the City. Curb Your Enthusiasm. 24. The Simpsons. The Daily Show. Frontline. My So-Called Life.*

Digital or analog, television is chiefly about entertainment—ask anyone at the networks or Netflix or Amazon. But the people who make it consider it art. Ask them about why they use no midrange shots, why they use nonsync sound at times, why they dislike golden palettes, or why they do or don't reloop. They will tell you this or that about Michael Mann, or even about Werner Herzog or Antonioni. These are filmmakers, and if you listen carefully you will discover they believe they are making art.

They will also tell you they dream of having a hit like *Dog the Bounty Hunter* on A&E.

And if you talk to the guy who made *The Hills*—one of the reality shows that bumped its stars to the covers of weekly tabloids, one of the greatest coarseners of culture—he will also tell you about palette, about sound, about cinematic references, about irony, about the representation of tears in digital video, and how to use subtitles for whispers.

And viewers, if you lurk around their message boards, will tell you they watch *Dog the Bounty Hunter*, *The Hills*, *Hoarders*, and *Extreme Couponers* for the same reasons they read novels. Because they have memorable, recognizable, and intriguing characters. Because they are beautiful. Because they inspire, provoke, amuse, soothe.

Just as BuzzFeed, Gawker, and *The Daily Show* have created new ways to disseminate news—to make what's important interesting and what's interesting important—HBO, the networks, Showtime, FX, and now Netflix and Amazon and Hulu constitute a remarkable climate to create new forms. I don't want to say it's a fertile climate, because it's the threat of Minow's wasteland that makes them work harder.

THE BINGE

From the minute Tony revealed his mob-life crisis on the pilot episode of *The Sopranos* on January 10, 1999, viewers marveled that television had become as good as cinema. That now seems like a strange response. *The Sopranos* was virtuoso work, and it looked lush, but it was considerably more than a mere movie. Movies run for, what, one hundred minutes? *The Sopranos* lasted for almost ninety hours. It takes years of devotion to savor a multilayered show like *The Sopranos*. Either that or a demented binge. You watch three episodes back to back. Or seven. A season. Three seasons. Soon you've eaten the whole bag.

In the past fifteen years TV viewers have fallen hard for the genre that Vincent Canby once dubbed the "megamovie." Season after season we get captivating, slow-burning, intellectually dense visual entertainment on cable and the Internet. Trained by *The Sopranos* in cable fandom, viewers now turn each of these shows into its own *Star Trek*, with a galaxy of detail-delirious fans. Megamovies are supersized entertainment watched with a magnifying glass.

But after *The Sopranos* we stopped decorously decanting our television, pouring out an oceanic series into hour-long or thirty-minute jugs, and doing our social drinking once a week.

Instead, using DVRs or cable and Internet on-demand services, we've been hoarding the shows for benders. Done alone, with a laptop or a tablet in bed, this kind of TV consumption really can resemble a drug-like binge, but the shows are so extraordinary that the pixel-snorting fests bring more pleasure than guilt. It's somehow more dignified than the "marathons" the old Viacom channels used to run of eye candy like *The Real World* or *Behind the Music*. During those marathons you subjected yourself to ads, previews, recaps, and all manner of repetitive moronism, usually because you couldn't be bothered to leave the couch.

With a binge you're less disdainful, more riveted. The shows are styled like Tolstoy, and promises abound that they reward close viewing. You don't turn vegetable watching them; you turn critic, leaning into a laptop, popping on Twitter, neurons firing, racing to make original connections and anticipate themes and plot. Or so it seems, anyway. Maybe the old marathon was more like a self-loathing drug, while the new binge on upscale dramas is like another, fancier drug. Maybe sherry. In any case today's binger certainly talks as though she's been to a film festival or studied deeply respectable, midlist novels when she's coming down from *Game of Thrones*.

There's also a new kind of intimacy possible when a friend and I hole up with a long-line narrative absorbed in perfect synchronicity. Some kind of whale-like mental hum seems to pervade a room when opening credits are rolling for the seventh *Homeland* in a row. (First season refrain: *I won't—I can't let that happen again.*) One person loves the credit sequence (that's me), while the other finds it misleading and spastic (that's Jamie). Every time we watch it we find more to agree and disagree about. For those of us who are bothered by the impossibility of reading a novel with a friend, synchronized binge TV is the next best thing.

In 2012, starting with *House of Cards*, Netflix became the first network to fully embrace this trend, releasing new TV series in Costco-size batches instead of one episode at a time. On July 11, 2015, thirteen brand-new episodes of *Orange Is the New Black* were laid out in defiance of portion control, like buttered rolls and creamed spinach at Western Sizzlin'. When *Orange* is in season Twitter seems to crackle with women just coming to after a nightlong bacchanal.

Binge-watching—complete with hangovers—only continues the conceit of decadence that's deep in the experience of all television, digital or analog. I long ago realized that neither streamable TV nor YouTube nor any kind of digital video should ever shake off its shady, bad-for-you reputation. Without it there would be no evolution, no fugitive glimpses of YouTube beauty, no robust digital cinema. Yes, cinema. But *shhh*, don't call it that.

LA NAUSÉE: VIRTUAL REALITY FINALLY EMERGES

Digital video in our time may have reached its apotheosis with virtual reality, which moves beyond fidelity, immersiveness, and 3D to something magisterially called "presence." Virtual reality, long considered a benighted and niche technology that would never find an audience, came into mainstream view in 2014, when Facebook acquired a company called Oculus for $2 billion.

I couldn't wait to try it, and I sought out a demo straight away. At the groovy downtown Manhattan offices of Relevent, a marketing agency that has created a virtual reality experience for HBO to help promote its hit series *Game of Thrones*, I walked into a steampunk cage the size of a phone booth, made of iron and wood. A kind of VR concierge then fitted me with headphones and the Oculus Rift, as the company's flagship product is called, a blocky set of black maxigoggles with an internal screen positioned inches from the eyes.

I promptly lost awareness of the screen, and after a few seconds a bass speaker under the floorboards began to boom. All I knew next was that I was shooting up, as in an outdoor elevator, to a windy summit and then trudging through lightly packed snow—crunch, crunch, crunch—onto a vertiginous ledge of ice.

I didn't turn my head. I felt paralyzed and choiceless, simultaneously propelled and enfeebled, as if I were being walked in a Baby Bjorn. Nervously laughing, I spoke aloud, shouting as if over wind, "I am in an office in Manhattan. Everything is fine. It is a workday!" I did this because I was fooled, profoundly fooled, and I needed to remind myself—and the PR team I half-remembered was watching me—that I wasn't six inches from extinction.

Immersive, transporting, revolutionary. But most of all not nauseating. That's the term that sets the Oculus Rift apart from the long line of demoralizing virtual reality Edsels that preceded it. The chief asset of the Rift—more than its dazzling specs, more than Facebook's sizable investment in it—is its dignified, nonemetic quality. All hail: the Oculus Rift doesn't make you vomit.

That's a crowning achievement. In 2012, when Palmer Luckey, Oculus's founder, presented his homemade VR headset at the New Frontier program at the Sundance Film Festival, I was eager to try it. After donning the DIY unit, which was held together with gaffer's tape and loaded up with "Hunger in Los Angeles," a haunting immersive-journalism project by a writer named Nonny de la Peña, I fell sick. Sick unto death, or so it seemed. The first thing that deserted me was interest in the spectacle; my own biological crisis monopolized my curiosity. The word *rift* thrummed in my swimming head. Uncanny illusions produced by the Oculus headset had indeed cleaved an unbridgeable rift between the evidence of my senses and an awareness of space and time deeper in my body.

This was not good. In seconds cognitive dissonance turned into something existential: bona fide Sartrian *nausée*. To hell with politeness. I ripped off my helmet and speed-walked, sheet white, past the art crowd at Sundance, in panicked search of a place to vomit. It took me several days to get my equilibrium back. And two years to try virtual reality again.

From the academic experiments and aerospace simulators of the 1960s to the Sega VR headset, Nintendo's Virtual Boy, and the Virtuality arcade games of the 1990s, virtual reality has always sounded fantastic in theory but felt in practice like brain poison. No wonder the progress of VR technology went more or less dark between 1998, when VR arcade games petered out, and 2012, when Oculus started to make Kickstarter rounds. Virtual reality was an abject failure right up to the moment it wasn't. In this way it has followed the course charted by a few other breakout technologies. They don't evolve in an iterative way, gradually gaining usefulness. Instead they seem hardly to advance at all, moving forward in fits and starts, through shame spirals and bankruptcies and hype and defensive crouches—until one day, in a sudden about-face, they utterly, totally win.

Virtual reality—the digital production, in a headset, of an immersive and convincing audiovisual illusion—is a dream that dates to at least 1968. That was the year Ivan Sutherland at MIT unveiled his "head-mounted display," which quickly earned the ominous nickname Sword of Damocles: a terrifying room-size VR machine, with a helmet so spine-crushingly heavy that it needed to be supported by a mechanical arm suspended from the ceiling. During the ensuing decades the military and NASA each seized on the VR concept in the hope of creating flight and combat simulations. But invariably these led to "sim sickness," a nausea so bad that it traumatized the people it was designed to train.

Forays into consumer VR hardly fared better. Jaron Lanier, the artist and computer scientist, popularized the term *virtual reality* in the mid-1980s, when he left Atari to sell VR goggles and gloves at the short-lived VPL Research. (VPL filed for bankruptcy in 1990.) Recreational virtual reality flamed out in the

1990s with a handful of unfun, overhyped, and physically sickening arcade games by a company called Virtuality. Developers like eMagin, Vuzix, and Nintendo still quietly plugged along, but the persistent nausea problem turned VR development into a grim, frustrating, even embarrassing business. Even just a year ago, if you asked most rank-and-file gamers about virtual reality, they might have said it was a nice old sci-fi idea but too expensive and far too stomach-churning to pursue seriously.

It's hard to remember, but the same skepticism once dogged the two devices that now define the sea change known as "mobile": the smartphone and the tablet. In the 1990s the Apple Newton, a tablet and "personal digital assistant," was considered a marvel by the specs. A single-surface networked personal computer, it came with an impressive screen and plenty of memory. But consumers rejected it. A handheld device that didn't fit in pockets, play music, display photos, or even make phone calls? The somber Newton didn't thrill anyone, so users didn't bother to make room for it in their everyday lives. Instead they roundly mocked it for its price and its many bugs. Merely debugging it seemed out of the question, though; its failure was taken as proof that nobody wanted a tablet computer. After the Newton's disgrace, Steve Jobs declared the tablet categorically discredited. "It turns out people want keyboards," he said in 2003.

And he seemed to be right—that is, until people couldn't care less about keyboards. Starting in 2007 with the iPhone, people spontaneously seemed to switch desires: now what they craved for their texting, emailing, social networking, and Web surfing were touch screens. The iPhone, literacy-defying device that it is, rejected the keyboard; the BlackBerry, of course, hung onto its physical keyboard all the way to oblivion. And by 2010 the keyboardless tablet made its reappearance, now called the iPad.

Similar reversals happened in the mid-2000s with ebooks and video calling, two long-dreamed-of technologies that appeared perennially hopeless until Skype and Amazon made them suddenly ubiquitous. And it seems to be happening with virtual reality today. Indeed fans of the Oculus Rift discover a pleasure so deep that John Carmack, Oculus's chief technology officer, invokes it with a particular solemnity. It's called "presence." To achieve presence with an Oculus headset means to be suffused with the conviction—a cellular conviction, both unimpeachable and too deep for words—that you are in another world.

Presence is still coming into a definition, but we know two things about it: it feels good, and it's different from verisimilitude. It is VR's equivalent of the "rapture" that cineastes used to describe being under the silver screen's spell in old movie theaters.

As Norman Chan explains on Tested.com, a virtual world decidedly does not feel like reality. For one, it still entails game graphics, rendered in a special way that schematically simulate depth. Even less true to life is the effect of personal disembodiment, as the user's own body is left largely unrepresented in the virtual world. But with presence, Chan explains, you do get a profound sensation of space, causing you to forget you're staring at a screen. Presence is fragile, but when achieved, it's so joyful and sustaining that those who touch it tend to fall silent.

I've found presence twice in Oculus experiences: in HBO's VR game Ascend the Wall and in a live-action airborne tour of a Dubai skyscraper. Each time it was glorious. The skyscraper tour came as part of a reel by Total Cinema 360, a production company in New York founded by two former NYU film school students that creates virtual reality programming as well as interactive, immersive omnidirectional video. The filmmakers seated me with the Oculus headset, on which I witnessed, taking place all around me, various cinematic vignettes: two lovers sharing pillow talk, a rock concert from the stage, and then, suddenly, the Dubai flight, which the company made as part of a recruitment package for a firm in the city.

In that flight I lost myself. I can't tell you how I became airborne, exactly; maybe I was in a harness, in a parachute mysteriously ascending. I sailed close to the sky-high architecture, somehow alongside it, where I could examine it, as if I were Philippe Petit on a high wire. But also entirely safe. I could look up at the sky and down—way down—without fear, as my wordless physiology signaled to me that I was rising and in zero danger of falling. This was flight, and not the nightmare kind with crashes: flight as in the best dreams of being winged and soaring, as happy as you've ever been.

No wonder users are clamoring to test the Oculus. Every time Ascend the Wall appears on its global rounds at *Game of Thrones* meetups, there are round-the-block lines of people wanting to try it. Some stay just to confer on their ice-wall experiences and watch others freak out during theirs. It's all pretty trippy, and the whole subject of virtual reality brings out the Timothy Leary–style psychonaut in those who enjoy it. Sometimes things truly get into Leary territory. A friend of mine told me she didn't need virtual reality because she'd recently tried DMT, a psychedelic compound known for producing powerful spiritual visions. I asked VR developers about DMT, and they nodded approvingly. "Oh, I like Oculus *and* drugs," one VR curator told me, as if to reassure me. Maybe I shouldn't have been surprised. This is a crowd that likes to hallucinate.

Even old hands at virtual reality generally submit to new experiences with a kind of trip sitter, a role borrowed from drug culture, in which people try hallucinogens with a sober partner who enables the enjoyment and ensures the safety of the one who is tripping. I myself have never tried VR without a trip sitter, and sometimes I think that reassuring nearness was half the satisfaction of the experience. It surprised me, actually, how intimate and touching I found it to have my Oculus headset fitted to me, to be given gentle cues about how to explore the various illusions, to be occasionally reassured that I wasn't going to disappear into a dream world, to have the headset removed when I looked bored or uneasy, and finally to be encouraged as I raved about various VR miracles.

To create and experience presence requires a keen sympathy between technology and neurology. Virtual reality sickness, most believe, is produced by a brutal conflict among sensory inputs.

Under the spell of VR the eyes and ears tell the brain one story, while deeper systems—including the endocrine system, which registers stress; the vestibular, which governs balance; and other proprioceptors, which make spatial sense of the body's position and exertions—contradict it. The sensory cacophony is so uncanny and extraterrestrial as to suggest to the organism a deadly threat.

If nausea is the body's dysphoric response to the uncanny, presence is the euphoric one. This is what most intrigues Oculus programmers. As the headset's hardware continues to improve—higher resolution, more frames per second, better positional tracking, and so on—Oculus programmers have kept alert to the idiosyncrasies of individual neurobiology. Many developers, felled by nausea in their first virtual reality forays but now safely acclimated with "VR legs," have a high tolerance for sensory incongruities. "You get a kind of immunity to V.R.-induced nausea," Eric Greenbaum, a Manhattan-based developer, told me in email. "It's a bit of an issue for V.R. development. Because developers often have a very high nausea threshold, they can't necessarily be a good judge of whether the experiences they are creating are going to be comfortable for the average user. It's one of the reasons that lots of user testing is so important."

In the fall of 2014 the tiny Stream gallery in the Bushwick neighborhood of Brooklyn, which exhibited an illusory gallery to be browsed entirely with a first-generation Oculus Rift developer's headset, employed docents with mops exclusively to clean up the real vomit produced by beholders overcome by the psychic dislocations of virtual reality. (For a bright young person looking to break into the Brooklyn art scene, this docent-mopper gig might be as avant-garde as it gets.)

But no one at Stream ever had to mop. Yes, many beholders of the intriguing VR artwork *Desktops*, by Terrell Davis, a sixteen-year-old Internet artist, did hit a neuro wall and stopped navigating around his angular virtual landscape, which incongruously crossed modern Dubai and Ancient Greece. I myself felt faint after a few minutes and quit. The curator, Kip Davis, nearly ripped the goggles from my skull when I said I was feeling woozy. ("We stop immediately," he said, "at the first sign of nausea.") Still, no one barfed. This is a meaningful achievement for a teenage artist working in virtual reality: his medium is no longer actually vomitous.

Neurological, technological, relational, psychological: there are so many moving parts in the creation of a full-dress illusory world. Virtual reality developers and fans regularly cite the "suspension of disbelief," a notion advanced two hundred years ago by the poet and aesthetic philosopher Samuel Taylor Coleridge. Of course Coleridge was talking about a reader willingly setting aside his skepticism about a story after being invited to do so by certain literary effects. An Oculus Rift experience, by contrast, involves something closer to a *forcible* suspension of disbelief. It's fitting, maybe, that the first theorist to use the phrase *virtual reality* seems to have been Antonin Artaud, in *The Theater and Its Double* (1938). Artaud's "theater of cruelty" aimed to expose spectators to the dangers of life, engulfing them in a tumultuous vortex that would leave them powerless and unable to escape.

In virtual reality the notion of powerlessness cuts two ways. Sometimes when I listened to developers talk about their eagerness to "immerse" audiences in multisensory experiences, I thought I detected a less savory desire to *imprison* them in programming, to leave them with no sensory exit. As much as nausea, it was that attitude that made me, sometimes, want to throw the Rift against a wall, to shatter that pricey digital alchemy, and gulp some open air, and with it the reassuring freedom and natural laws of real physical space. Swaddled in goggles and headphones, your power-forward senses (sight and hearing) are steamrollered by a visually and aurally complete universe designed precisely to seal out opportunities for doubt. Virtual reality sickness, *la nausée*, can be seen as the body's radical disbelief in this illusion. It surfaces to remind you, in horror, of your subjectivity and to force you to reclaim your sensory autonomy.

Oculus programmers often talk, as Coleridge's fellow poets might have talked, about how to keep audiences under their spell. Carmack, the Oculus CTO, has discovered some neat tricks to circumvent the "sensitivities" that some users have to VR phenomena, like the smearing, ghosting, or flickering in VR images. He has highly technical strategies for "deghosting" images, as well as for achieving "submillimeter accuracy with no jitter" in the Rift's positional tracking. As for the flicker, Carmack advises keeping the periphery of a spectacle black to calm the flicker-averse. (He praises something called Oculus Cinema: "It winds up being a dark experience at the corners, so it's comfortable from a flicker standpoint.") In our gentler commercial idiom, used for movies and games, in which escape from reality is considered a given good, this comfort is that thing called presence.

Even if Imax 3D movies make you queasy, you might be able to stomach the Oculus Rift. The company's shiniest prototype, which it internally calls Crescent Bay—with integrated audio, no visible pixels, better tracking, and a light, ski goggle–style headset—strikes many who have tested it as close to perfect. But even a year after the acquisition Facebook and Oculus were determined to keep the technology out of consumers' hands until it was blissful and bound to be world-historical, on the scale of the iPhone.

"Every 10 to 15 years a new major computing platform arrives," Mark Zuckerberg said in a 2014 earnings call for Facebook. "We think that virtual and augmented reality are important parts of this upcoming next platform." Zuckerberg then threw down the gauntlet in his insouciant way. The Rift, he said, "needs to reach a very large scale, 50 million to 100 million units, before it'll really be a very meaningful thing as a computing platform. . . . That'll take a few cycles of the device to get there, and that's kind of what I'm talking about. And then when you get to that scale, that's when it starts to be interesting as a business in terms of developing out the ecosystem."

Marketers, educators, scientists, and, of course, gamers are already imagining an internal ecosystem for virtual reality. Armchair travel. Risk-free skydiving and zip-lining. Gender-bending with virtual bodies. Classrooms of avatars convened with people all over the world. Surgical demos. Virtual hikes in the Andes and sprints on Fiji beaches. But whatever its "use" might be, VR is not fundamentally a pragmatic technology, which is why it begins with gamers. If it works, if it catches on, it must first give pleasure—and be fun.

It's curious that James Cameron, a director known for his embrace of technology in the name of cinematic spectacles, was the first of many to dismiss Oculus as "a yawn." Cameron's pose brought to mind something Ian Cleary, whose firm designed Ascend the Wall, told me. I'd asked Cleary if anyone visiting the snow-capped virtual land of the Seven Kingdoms from *Game of Thrones* is ever able to maintain an attitude of boredom and indifference. He thought back to one or two guys he had encountered on the exhibit's international tour who endeavored mightily to play it cool during the ice-wall ascent. "They seemed worried that everyone is looking at them, and they are determined not to react, but there are overreactors too," he said, alluding to an actress who tried Ascend the Wall and turned squeally and hysterical. Both reactions can seem contrived and engender nostalgia for proportional responses to a three-dimensional world of energy and matter that behaves in predictable and ancient ways, in harmony with our capacity to appreciate it.

"The truth is," Cleary said, "virtual reality just creates a deep hunger for real-world experiences."

5

MUSIC

THE WHEAT PLAYED BACH

My delight and then discontent with music on the iPod inspired me to think closely about what is lost and what is gained on the Internet. Digital songs slammed me into the magic of the digital revolution. Then, precipitously, those same compressed chords brought the loss.

"I have told you this to make you grieve," Dante wrote. Maybe that's why the likewise labile romantic Steve Jobs—who once, high on acid, conducted a wheat field playing Bach—furnished us with all that digital music in the decade before he died. His iTunes shot billions of songs straight into our limbic system precisely to *make us grieve*. Just as it's impossible to write about Western poetics without close attention to the language of Dante, so it's impossible to consider the aesthetics of the Internet without close attention to the soundscape of Jobs and Apple.

But first: I loved my iPod, which I bought soon after it appeared from Apple, at the very end of 2001, as the affronted nation roused itself for war. To my own frustration, I had failed to master the gray market peer-to-peer file-sharing services, such as Napster and Limewire, that I knew offered music in the Internet's anarchic spirit. The songs stuttered as they downloaded, which took ages; I never seemed to end up with the right version. I was thirty, tired, working to cover rent. I wanted music to be simple and legal and to work. So I turned to what the writer Brianna Snyder calls "the nanny-state iPod." On it, music was expensive, uncool, and safe—and no longer a sport only for hackers, teens, or postpunk muso-kings. But neither did you have to be a dupe of the music business frequenting the new "listening stations" at Tower Records. Now a person like me, whose most profound experiences of recorded music had come with mixed tapes, my roommates' CDs, and overheard passages of AM radio, could make a good-enough playlist and sail through my subway commute with a skullful of Norah Jones.

It was always Norah Jones. Jobs and his team had included CDs of Ella Fitzgerald, Sarah McLachlan, and Alanis Morissette (and many other male artists) with the first iPod package for tech critics. But Jones wasn't just incidentally compatible with the iPod. Her music seemed *made* for the iPod. Music critics argued that the anodyne music of *Come Away with Me* (2002) suited iPod owners because the expensive gadget announced its fans as comfortable and bourgeois, and Jones's drowsy songs politely preserved that comfort. That was mean. But maybe true. With little guilt I drank in the nondisturbance, which drowned out the black helicopters that swarmed around New York Harbor after 9/11, as well as the filibustering of hawkish pundits gunning for attacks on Tora Bora or Baghdad.

When the iPod was first released I was recovering from an extensive second surgery on my left mastoid to remove a cholesteatoma, a destructive but noncancerous tumor in my skull bone. Sometimes I used only one earbud. As my ear recovered and I examined the many CT scans and MRIs that showed my disease, I was acutely aware of the small, fragile membrane that stood between my brain and the world. The membrane had ruptured—*perfed*, in the surgeon's idiom—but it was now fixed, and I wanted to test how well the new eardrum played. The simultaneous threat and promise of deafness and silence was also on my mind.

I don't know why I didn't come. I listened and listened and listened to Norah Jones, and with my discreet iPod display and personal wheel no one was the wiser. I didn't have to answer to anyone for my timid choice. If I'm being honest with myself, what I wanted from music in the early 2000s was convenience, pleasantness, and zero possibility of shame or anger. Andrew Sullivan, the consummate cultural elegist, could have been describing me in an article about America in the online London *Times* in 2005:

Nightlife is pretty much dead . . . but daylife—that insane mishmash of yells, chatter, clatter, hustle and chutzpah that makes New York the urban equivalent of methamphetamine—was also a little different. It was quieter. Walk through any airport in the United States these days and you will see person after person gliding through the social ether as if on autopilot. Get on a subway and you're surrounded by a bunch of Stepford commuters staring into mid-space as if anaesthetised by technology. Don't ask, don't tell, don't overhear, don't observe. Just tune in and tune out. It wouldn't be so worrying if it weren't part of something even bigger. Americans are beginning to narrow their lives.

Anaesthetized, yes. Stepford, if you like. But *narrow*? My life didn't feel narrow. Compressed, more like. I was beginning to compress myself. Though on one level I was gliding through train stations listening to the ethereal lullabies of Ravi Shankar's beautiful daughter, on another level I was craning to hear the mystery of digital compression and sounds that seemed to be otherworldly, genetically modified.

Like the indigenous people in Adele Horne's film *The Tailenders*, who quickly convert to Christianity after exposure to the miracle of disembodied sound, sound from the iPod was converting me to something. Television sets were still analog in those days, as were most phone calls. But the iPod, shooting bit by bit into my very brain, was digitizing me.

In 1995 in *Being Digital*, Nicholas Negroponte had enjoined readers to embrace our status as information bits rather than atoms of matter. Sure enough, I was somehow *becoming* the thing I was studying. This was knowledge through what Thomas Aquinas had called, in the context of divine wisdom, "connaturality"—the sharing of a nature with another. To me this was the iPod's magic: it transformed me and made me digital.

The music that we hear on mobile devices is not music, exactly, but a *representation* of music, in bits. Like other representational arts—realist painting, journalism, photography, film—MP3 music is an extremely persuasive and pleasurable illusion. The MP3 representation is so seductive, in fact, that we regularly take it for the thing itself. We're like the mythical birds said to have pecked in vain at Zeuxis's fifth century BC trompe l'oeil painting of grapes, with their illusion of volume. We mistake the painting for the fruit. No one doubts that she's listening to music when she listens to the paintings of music available on her laptop or mobile phone.

The pervasiveness of MP3 music is yet another way that the Internet, broadly conceived, is a massive work of *representational* art, populated by our online avatars and increasingly recruited to stand in for reality itself.

But what *is* MP3? For years I assumed that the "MP" stood for "microprocessor." It does not; it's derived from the acronym for the Motion Picture Experts Group, a high court of authorities in the engineering of audio and video. First convened with a few dozen members in 1988, the MPEG—now 350 members strong—still sets global standards for the compression, decompression, processing, and coding of representations of video, audio, and video-audio combinations. In 1993 the group encoded movies and audio for the bit rate of CDs, their first project. That first code, in which the video looked primitive, was named after the panel itself: MPEG-1. A subset of MPEG-1 was the audio, standardized by the so-called Audio Group (a chairman overseeing members from fourteen research institutions), which they divided into layers. Of these, MPEG-I, Audio Layer III—shortened to MP3—was arguably the group's biggest success, though only when young people found it online and started seeking the codecs that "made your computer sound like a stereo," as Steven Levy puts it in *The Perfect Thing*.

From the start MP3 pulled off two miracles. First, it dramatically reduced the amount of data required to represent an audio signal. Picture an artist who can conjure Abraham Lincoln's face with a few lines instead of needing to create a dense and expensive oil painting. Second, the representation could really pass: *it sounded like the original uncompressed audio!* Or so the panel of engineering experts concluded.

The data that the engineer-authors behind the engaging textbook *Digital Compression for Multimedia* call "important signals"—voice, music, TV, and movies, the stuff we pay to see and hear—could now be represented in bits with ferocious fidelity *and* ice-cold efficiency. That was the magic that even I, the end user, could sense in my earbuds.

I found that a big part of the magic of the iPod was also that the invisible encoded sounds coolly defied the material reality of music. As a reader I like symbolic abstraction, and in those days I craved more than usual the break abstraction offers from the material world. After the evaporation before our eyes of the Twin Towers and thousands of people less than a mile from my apartment into poisonous air thick with legal-paper confetti and fumes, the world of three dimensions seemed treacherous. So I appreciated the iPod's flat-out rejection of music's strings and reeds, as well as the clack and crack of horrid CD jewel cases. I cherished the break from real life. As Levy writes, "Vast swaths of humanity . . . separate themselves from the bonds of reality" with the iPod.

Only later would I come to feel, in the prickly stereocilia of my inner ear, or maybe somewhere even deeper, something uncanny. This almost-music was missing something, something vital. It was, in the engineer's evocative jargon, "lossy." Something was lost in the digitization of music, and as much as I wanted to revel in the magical pleasures of the iPod I needed to take the measure of this loss. The magic, and the loss.

Part of the pleasure around music that Apple made available with iTunes was not disembodied sound but intense synesthesia, a hallucinatory reworking of Jobs's melodious wheat fields called the Visualizer. This program, which, as it's not for all tastes, was never immediately obvious in the iTunes interface, generates abstract light shows you can elect to watch while you play music on your computer. Its luminous tableaus move with the sounds, sometimes in tight synchronicity and sometimes like a lazy dancer who only sways or shuffles his feet. For quite some time I spent album after album marveling at the protean scenes on the Visualizer, trying to fathom the beauty of the program and divine its crazy logic.

The Visualizer is something that either fascinates or slightly bugs you. When I bring it up in conversation, I'm amazed that many people twirl their fingers and goggle their eyes to imply it's just a psychedelic thing. Sure, there are some 1970s-style starbursts on the Visualizer, but there's also extraordinarily inventive video art that far surpasses mystical kitsch. By turns, the strange forms on the Visualizer envision the music, comment on it, laugh at it, and succumb to it.

Most self-described synesthetes, who interpret sensory experiences with more than one sense (and who are predominantly female), attach color to musical timbres or tones. Not being a synesthete myself, I try to follow closely the colors on Visualizer, those incandescent and surprising half-tones that I didn't realize my monitor had in its chromatic arsenal. They're beautiful.

But the Visualizer offers much more than color. There's shape and form too. The terms *low* and *high*, characterizing notes, have spatial implications, as do *deep* and *soaring*; that's not lost on the minds behind the Visualizer. As with its palette, its shapes are mesmerizing: paisley, oil spills, blue smoke, embraced-by-the-light tunnels, electrocardiograms, icicles, video-arcade spaceships, amoebas, flames, flares, static, smears of headlights at night, half-developed Polaroids, Cy Twombly scribbles, viscous red ribbons that belong in an ad for Cognac. What *is* all of this? Fleeting and intimate contemporary forms from medicine, nature, video games, art, movie clichés, technology, advertising, anime. Amorphous objects that come right up to the brink of being representational, iconic, or logoish—as if tempting the copyright police—only to dissolve or metamorphose.

Rapid transmogrification keeps you from trusting your eyes, but rather than being frustrated, I feel liberated. I turned on the singer-songwriter Laura Veirs's album *Year of Meteors* while the Visualizer played, and my eyes, not charged with making sense of the objects before them, seemed to cede their role as meaning-giving sense organ to my ears. I have heard that album's intricacies so exactly now—the Visualizer never represents a song the same way twice—that I feel as if I have tasted its grooves, slept in it, and inhaled every note of every song.

If Jobsian sense-crossing sounds too far out, consider that digitization has sifted images and sounds into bits: they are one and the same. And at every turn (Tidal, Vine, Snapchat) the temptation to turn a whim into an obsession is hard to resist. The only act that's impossible is consuming art the old way: treating the Internet like a record collection that might be dipped into with a balance of equanimity and curiosity. We're officially through the looking glass, everyone; we might as well stop to smell the music and hear the new air.

THE ANTI-EDISON

Though preceded by other MP3 players and dozens of mobile phones, the iPod is still the Adam of handheld smart technology. Its formidable DNA makes itself known on our phones, no matter the brand. It's worth noting that our current everything-gadgets (smartphones) did not begin as compasses, calculators, calendars, clocks, or communication devices. They began as music boxes. In the beginning was the desire to listen to music.

Unlike Edison, who conceived the phonograph as a dictation device for men at desks, declaring that its use for music was a passing fad, Jobs made music a priority with his invention. This makes sense. Edison had a well-known fetish for labor, communications, and exertion for its own sake. Jobs styled himself as an atelier type, a seeker who considered himself, at heart, as he told Walter Isaacson, "a humanities person."

True to the cultural ideal of his time, Edison fashioned himself as a workhorse experimenter, while Jobs, who grew up in northern California in the 1960s, was drawn to a set of cultural practices that had ecstasy, self-love, and truancy as their master values. Jobs was also born into a world in which consumer tech was defined by "the high-fidelity boys," as Meyer Berger described them in the *New York Times Magazine* in 1953. This clique included "the hot-eyed and intemperate fanatic whose chief pursuit is not music but extremes in sound—the lowest booming bass; the highest biting note, tremblingly caught before it takes off for infinity."

Jobs was a hi-fi boy from the start, drawn to sonic infinity. In fact his first collaboration with Steve Wozniak had nothing to do with making computers; they were acquiring illegal Bob Dylan bootlegs and then finding clever ways to store the music (on reel-to-reel) and listen intimately and obsessively to it (through headphones). The project prefigured the engineering and aesthetic of the iPod. The other thing they partnered to do early on was phone phreaking: matching certain AT&T tones and chirps so they could game the phone company and place free calls. Many phone phreaks—predecessors of today's trolls— prided themselves on perfect pitch. They were only the most cartoonish example of sound connoisseurs who prized fidelity over musicality in the sound signal.

Though Isaacson tries valiantly to impute to Jobs a love of science, Jobs himself remembers a childhood defined instead by Lutheranism, Zen, class cutting, poetry, literature, music, drugs, Barnumism, and antiauthoritarianism. Though he's said to have audited a physics class while he worked at Atari, the truth is Jobs virtually never concerned himself with science—with cells and atoms. Instead he was a technologist, drawn to bits. Technology and science share very little methodology, practice, or even assumptions. Like most technologists, Jobs experienced his interest in technology as an extension of his interest in art and the humanities. Unsurprisingly he gravitated toward only the trippiest tech: lasers, wireless amplification, feedback loops, and devices that could conjure illusions in sound and light. He thought the most exciting uses of technology were not the most functional but the most hallucinatory.

This made sense. In 1972, a disturbed young hippie freshly in love, Jobs took his girlfriend to drop acid in Sunnyvale, California. High, he discovered that rapturous synesthesia: the wheat field playing Bach, with him conducting. "It was the most wonderful feeling of my life," he told Isaacson. Music supplied by the brain and cathected onto tuneless things: plants, lasers, silicon. This was the splendid hallucination that would make the iPod, when it appeared, as good as revealed religion.

The difference between the phonograph and the iPod—analog and digital music—might be further explained this way: Edison's cognitive pleasures were derived from *electricity*, with its on-off binarism and its passive components (relays, switches, etc.). In contrast, Jobs was turned on by *electronics*, which permit and channel the flow of electrons in unpredictable, rococo, and non-linear ways. Electricity is the province of the engineer and the rationalist, founded on inviolate binaries. Electronics, which is built on the semiconductor (the silicon itself, which is neither conductor nor insulation but somehow both), is the province of the irrationalist, the deconstructionist, the druggie, and the mystic.

What does this have to do with music? Quite a bit, in fact. Edison's music box, the phonograph, was billed as a dictation device that would make the workplace more efficient. Jobs's music box, the iPod, was designed to make us *less* efficient—to supply each mind with literally its own drumbeat, Thoreau-style, and to drown out the voices of those who would give us orders. Allow us, in other words, to think different.

I certainly used it that way, well into the 2000s. I loved downloading all those songs, and in the indolent never-ending voice of Norah Jones I heard the injunction to take my time, tune in, hustle less, and not let others boss me around.

So that was the magic. The iPod fulfilled certain long-standing desires for connection to the music of existence, and the musical notes it piped into our ears and heads were like the wine on the lips, a signal that pleasure would soon flood our bloodstream. Very quickly we developed a Pavlovian response to that iPod taste and began to crave it.

The intoxication couldn't last, though. A rapturous hour of isolation (in public) with the iPod soon gave way to a kind of nausea, and complaints surfaced among iPod users that whatever we were listening to in our earbuds, it wasn't quite music. But why? It was hard to place the discomfort. Maybe, some old-time music fans said, it wasn't music because it was no longer "album-driven," as purchases on iTunes were generally by single songs, and thus didn't require sustained listening. The shuffle feature, which delighted Steven Levy and other DIY DJs, wrested songs from their album context and violated the integrity of a studio creation. Or maybe it wasn't music because it was now way too far from the hi-fi fantasia we'd grown up with: a father in giant headphones, plugged into a stereo topped with a gently spinning LP on a turntable. *In the groove.* Was it even possible to get in a groove when the music—those invisible MP3 files—lacked grooves, strings, keys, needles, and everything else three-dimensional that can be said to "make music"? Jobs himself had recast "the groove in the record" from a psychic sweet spot to a place to get mentally stuck, as he told David Sheff in *Playboy* in 1985. But maybe moving out of the groove just meant moving on from music entirely. Or maybe it wasn't music because the iconic earbuds that came with the iPod wouldn't stay put or were killing our hearing.

All of these explanations for the melancholy induced by

music on the iPod are valuable. Most earlier discussions of the transformation of music by digital technology had focused on the production end: MIDI and synthesizers and digital amps. The iPod convinced me that music is transformed and potentially diminished on the consumer end too.

Take those once iconic earbuds—right up next to the ear, as close to the body as it gets with music. All headphones are packed with technology. When an audio current passes through the device's voice coil, it creates an alternating magnetic field that moves a stiff, light diaphragm. This produces sound. If you think about all the recordings, production tricks, conversions, and compressions required to turn human voices and acoustic instruments into MP3 signals, and *then* add the coil-magnet-diaphragm magic in our headphones, it really is amazing that the intensely engineered frankensounds that hit our eardrums when we listen on iPods and iPhones are still called music.

The hazards of headphones can't be overlooked. A decade after the iPod's appearance—and as mobile devices with earbuds seemed to be reaching full population saturation—the *Journal of the American Medical Association* reported that one in five teenagers could no longer hear rustles, whispers, or the plink of raindrops. Many researchers attributed this hearing loss to exposure to sound played loudly and regularly through headphones. Earbuds in particular don't cancel as much noise from outside as do headphones that rest on or around the ear, so earbud users typically listen at higher volume (as I did) to drown out interference.

Headphones were a big Jobs favorite, and there would be

no digital music without them. At the same time, the history of headphones has always been one of unexpected uses and equally unexpected consequences. They were invented more than a century ago, the brainchild of Nathaniel Baldwin, a tinkerer and mystic in the Jobs mold. A Utah Mormon, Baldwin grew frustrated when he couldn't hear sermons over the noise of the crowds at the vast Salt Lake Tabernacle. (Today's MP3 listeners are almost certainly choosing to tune such things out.) Baldwin first designed an amplifier, then added two sound receivers connected by an operator's headband. According to legend, within each earphone was a *mile* of coiled copper wiring and a mica diaphragm to register the wire's signals with vibrations. When the U.S. Navy put in an order for one hundred Baldy Phones in 1910, Baldwin abandoned his kitchen workbench, hastily opened a factory, and built the prosperous Baldwin Radio Company. His innovations were the basis of "sound-powered" telephones, or phones that required no electricity, which were used during World War II.

It's not incidental that Baldwin first imagined headphones as a way to block out crowd noise and hear sermons. Workers and soldiers have long used them to mute the din of machinery or artillery while receiving one-way orders from someone with a microphone. From the beginning headphones were a technology of submission (to commands) and denial (of commotion). When World War II ended, submission-and-denial was exactly what returning veterans craved when they found themselves surrounded by the clamor and demands of the open-plan family rooms of the postwar suburbs. By then they knew what device provided it. In the 1950s, when Jobs was a boy, John C. Koss invented a set of stereo headphones designed explicitly for personal music consumption. In that decade, according to Keir Keightley, a professor of media studies at the University of Western Ontario, middle-class men began shutting out their families by using giant headphones and hi-fi equipment that re-called wartime sonar systems. When Sony's Walkman appeared in 1979, headphones became part of a walking outfit. Now head-phones and earbuds are used with MP3 players, mobile phones, tablet computers, and laptops.

Whatever you call it, we're listening to *something* when we use private music boxes—though "listening" is too limited a concept for all that headphones allow us to do. Indeed the device seems to solve a real problem by simultaneously letting us have private auditory experiences and keeping shared spaces quiet. But the downside too is obvious: it's antisocial. As Llewellyn Hinkes-Jones wrote in the *Atlantic*, "The shared experience of listening with others is not unlike the cultural rituals of commu-nal eating. Music may not have the primal necessity of food, but it is something people commonly ingest together."

Headphones in smartphones work best for people who need or want to hear one sound story and no other, who don't want to have to choose which sounds to listen to and which to ignore, and who don't want their own sounds overheard. Under these circumstances, headphones are extremely useful—and necessary for sound professionals, like intelligence and radio workers—but it's a strange fact of our times that this rarefied experience of sound has become so common and widespread. In the name of living a sensory life, it's worth letting sounds exist in their audio habitat more often, even if that means contending with interruptions and background sound.

Jobs was well-known for his abstemiousness and his use of diet and spiritual regimes to deny, transcend, and stifle the body's many appetites and tendency to decay. In his younger years the Buddha himself, whom Jobs, a student of Zen, must have wondered at, had done the same. But when the Buddha came to his enlightenment, he gave up starving his body and striving to transcend his humanness. Instead he accepted that only in absolute awareness of his mortal body—sensitive to the irreducible fact that each of us is a "dying animal," as Yeats put it—can connection with the deathless be made.

To those who love it music is often thought to contain an intimation of deathlessness, of immortality. Well before iPods, dreamy types went around with chords, notes, and phrases in their mind, open to the possibility that an audibly deteriorating body in space—a guitar made of ash wood, say, or a tenor made of flesh and blood—might produce something divine. In *this* realization Marcel Proust cites love's victory over death. In *Swann in Love* he writes of the persistence in memory of beloved phrases of music, "We shall perish, but we have as hostages these divine captives who will follow and share our fate. And death in their company is somehow less bitter, less inglorious, perhaps even less probable."

Music can offset death. That's what we're all saying, isn't it? As Jobs said of the iPod after September 11, "Hopefully it will bring a little joy to people."

Yet in the name of being fast, portable, cheap, and extensive, digital music forfeits depth. Just as a photograph of a painting, however dazzlingly ultra-high-def, can't convey the density of paint, of paint under paint, of canvas under paint under paint, neither does the MP3, with bits that paraphrase a piece of music, suggest the echo of the chirp of the bassist's sneakers on the wooden stage as he nervously kicks his foot or the sound of the backup singer's lungs still metabolizing pot smoke. For that multidimensional soundscape, you need to hear music live.

MP3 compression is predicated on the idea that one slice of data, skimmed off the top, can communicate a sound made in time and space by multiple bodies, collisions, textures, and movements. Like Zeuxis's realist paintings, none of which survives except in the legend of how they fooled birds, MP3 tricks the senses. It suggests volume with hints of sonic light and shadow the same way the first Mac interface was able to use graphic shading to suggest a stack of documents in two dimensions. We try to quench our thirst for music on MP3 but end up pecking at painted grapes.

The legend of the painter Zeuxis, however, does not end with the duped birds. Pliny the Elder recounts in *Naturalis Historia* that Zeuxis's painting of grapes was only one entry in a competition with Parrhasius, another painter in ancient Greece. Thrilled by the birds' reaction to his grapes, Zeuxis turned to Parrhasius and asked his rival to pull back the veil on *his* painting.

Not so fast! The veil, it seemed, was part of the painting, was itself an illusion. Gallantly Zeuxis (in Pliny's telling) conceded, "I have deceived the birds, but Parrhasius has deceived Zeuxis."

The lovely parable used to be cited in discussions only of volumetric illusions. But the supremely French psychoanalyst Jacques Lacan, a wonderfully trippy 1960s thinker in the Jobs mode, saw something else in it. Birds may be drawn to a spectacle, but humans are drawn to what's hidden. We refuse, as Lacan puts it in *The Four Fundamental Concepts of Psychoanalysis*, to have our gaze (or hearing) entirely *tamed*, to perceive only what an artist or auteur or musician wants us to perceive. We want to look around corners, probe mere paraphrase, demand what's being withheld. *Where* are *all the songs in this tin-can iPhone? Where is the bow, the groove, the drum skin, the fret, the human larynx?* Something is being concealed in these tiny, inviolate phones. We long to gain, then, what has been conspicuously squeezed out of "lossy compression."

That longing builds—in my neurons, I swear—as I listen to music from iTunes, these days mostly Neko Case and Vampire Weekend. *What am I missing, what am I missing?*

The possibility that others feel the same seems to find evidence in the entirely unpredicted renaissance in live music that has so far defined this century in music every bit as much as or more than the iPod. In live stage shows are errors, false notes, muffling, pauses, heaving breath, hollow feedback, rampant distortion, and all the familiar madness of uncompressed social space. Yet to everyone's surprise this past decade has become a bonanza for touring. By 2015, touring bands were generating some $20 billion for live shows, more than they ever had in history. "And that brings us to one of the biggest advantages of live gigs: scarcity," wrote Paul Resnikoff in *Digital Music News*. "In the end, a concert can't be instantly copied and duplicated, and neither can the social, in-person aspects that come with it." Epiphany.

NOSTALGIA FOR RANGE

In 2007, I at last met Funtwo, Jeong-Hyun Lim, the mesmeric musician I had watched over and over on YouTube. We sat down for a burger in Union Square. Out of nowhere he asked me a jarring question. "Why are the only people interested in shred guitar other Koreans? Why do Europeans and Americans only like Green Day and punk and not the complex digital stuff like Dream Theater?"

I said nothing for a minute, and then Funtwo supplied an epochal answer to his own question. "It's because Europeans and Americans are relaxed," he said. "They like to listen to music. They go to concerts. In Korea, we don't like to listen to music because we like to play music. So we get better and better at technique, and American musicians get better and better at taking risks with music that is not good."

Suddenly, I saw it—music in the American consumption economy (with more listeners than makers) versus music in the Asian production economy (with more makers than listeners). Of course! When consumers outnumber producers, technical quality in U.S. pop goes down, while the reverse is true in Funtwo's Korea. Even if Funtwo's confidence in American "relaxedness" were misplaced, and even if he didn't grasp how many Americans loved his style of playing, he was right that many of us had started to seek music that didn't promise perfection, but defied it—since "perfection" was identified with MP3s and other compressed sounds. Green Day fans may have been seeking music that sounded relaxed itself, flawed, somehow less determined to somehow win.

I never turned to Green Day, but I certainly developed nostalgia for a whole range of messy sounds I missed in the digital age. Certain music fans I knew were buying live-music tickets and vinyl records again for their "warmth"—what Mark Richardson at Pitchfork reduces to distortion caused by filtering and processing the bass for vinyl—but I didn't want vinyl. I wanted my analog phone back.

I don't know why the telephone, the analog landline telephone, was never formally mourned. What a many-splendored experience it once was to talk on the phone. You'd dial a number, rarely more than seven digits, typically known by heart and fingers. You'd refrain from calling after 9 p.m. or during dinner; there were many ideas of politeness around phones, and those ideas helped people pretend that the emotional chaos fostered by all that ungoverned, nonpresentational, mouth-to-ear speech—like whispering across great distances—didn't exist.

You'd endure the long brrrings with a pleasant stirring of nerves, a little stage fright. As many as ten, to give the household a chance to rally. On "Hello?" you'd identify yourself and ask for the person whose voice in your ear you, having waited, now profoundly desired. In the absence of the grammatical spasm "This is she," you'd learn whether your friend was in or out or somewhere in between (weird parents sometimes said "indisposed"), while your patience was casually requested. ("Hold on a sec. She's in the den.") You'd express thanks for the answerer's good offices. More waiting. Offstage noise. *Voilà*. Up would come the voice. Lucid, expressive, a perfect sonic spectrum; in those days a friend's phone voice was as much the friend as was her body in physical space.

A conversation could last hours upon dazed hours as you sat on your parents' bed, twirling the curly cord, or hauled the house phone into the bathroom, the better to monopolize family telecommunications. Chortling, gasping, sighing, sobbing, throat catching, or forming word after idle or impassioned word: you made every sound that humans make and thus joined your solitudes.

Intrusions came from others who wanted to use a household's sole line. Arguments could erupt, aggravated by how eager you were to conceal from your friend the noise of family discord or stern paternalism. Later you might learn you had been responsible for a busy signal encountered over and over by an important adult trying in vain to get through. The remote but ever-present possibility that you were creating real-world disasters, or in any case preventing their resolution, with your sweet nothings and mutual respiration deepened the pleasure of those long, desultory calls.

While they cannot be said to have abetted the swift completion of anyone's appointed rounds, the old phones—wireful phones, defined by the strong visible insulated copper circuits that crosshatched the land—came to be indispensable to anyone who longed for a complex social and emotional and aesthetic life, a reliable vocal-auditory miracle, intimacy, friendship, romance, furious down-slammings, hissed interruptions, and the awesomely strange sensation, via the mouth- and earpieces, of being inside someone else's accent, intonations, and sighs, ear canal and larynx and lungs. Your phone voice and manner were you. You thought a great deal about people who rhythmically and mysteriously inhaled and exhaled cigarette smoke while they talked or left long silences or didn't hang up immediately after saying goodbye.

Before voicemail, there were fears that call-borne opportunities might be missed forever, but there was no "We have a bad connection," "I'm going into a tunnel," "My battery's dying," "I have to take this," or "I have only one bar." In movies from the 1940s and 1950s people fight telephone static and encounter feckless operators and crossed wires; that maddening I-can't-quite-reach-you effect is something like what we have again with cellular and digital telephony. Sound signals, so unfaithful to the original they hardly seem to count as reproductions, come through shallow. You can hardly recognize voices. Fragile, fleeting connections shatter in the wind. You don't know when to talk and when to pause; voices overlap unpleasantly. You no longer have the luxury of listening for over- and undertones; you listen only for content. Calls have become transactional, not expressive. The oddly popular option to use the speakerphone means that you never know when what's left of the old telephone intimacy might be compromised. You certainly can't trust that it will be there anymore, ever.

Intimacy has flourished in other places, of course. There's

ingenuity and thrill to the pace and humor of texting, and email, message boards, and instant messages can be as emotionally rococo as the best of the old, gone-forever phone calls, which were written only on air. We haven't lost intimacy. We have lost only telephones.

When some teenagers were lying on their bed listening to music, enveloped in the warm sound of vinyl records, others were listening to friends, rapt at long-playing spoken-word music mixed just for the listener. Long phone calls were supposed to be a girly addiction, but those calls of the 1970s and 1980s were the only way to court girls, so boys too learned the art of them. Audiophiles now miss vinyl, and I miss those calls.

SONIC CY TWOMBLY PAINTING

Digitization has seen the rise of increasingly precise sound design, which, like MP3 music, doubles as commentary on the relationship among atoms, cells, and bits. In particular the best sound design for the movies has made an effort to supply what was missing from compressed music and cellular telephony: the body, the breath, and the inevitability of death and decay.

One formidable and ambitious soundscape that remains state of the art is Kathryn Bigelow's *The Hurt Locker*. It's a bomb movie that mutes its booms. It derives suspense by withholding the expected "boomala, boomala," as an Iraqi kid says while taunting an American bomb squad soldier about the "cool" soundtrack of Hollywood war.

The Hurt Locker is not cool. It is not a 1960s McLuhan set of tracks or a track on iTunes, aloof and clean. Instead it's hot and dry, a heaving desert parable with a mounting sandstorm howl at the center. The internal explosions matter more than the fireworks. Explaining the dynamics of roadside bombs in Iraq, Paul N. J. Ottosson, the film's supervising sound editor, told *Variety*, "You die not from shrapnel but the expanding air that blows up your lungs."

The top notes in the soundtrack are arid metallic clicks, snips, squeaks, and creaks, the chatter of wrenches and wire clippers, as bombs are defused in air itself so parched as to seem combustible. Men can hardly summon the spit or breath to speak. Much of the dialogue, almost all of which was recorded on location in Jordan (not looped in a studio), is delivered in headsets, as soldiers hiss into one another's helmets across desert expanses. To listen is to enter machinery, rib cages, ear canals, and troubled lungs.

For its cerebral, abstract, and still deeply romantic sound tableau, a kind of sonic Cy Twombly painting, *The Hurt Locker* won that year's sound-editing Oscar. Ottosson's alignment of death and silence, instead of death and booms, partakes of an aesthetic based on the idea that you're deaf when you die. Worth a listen, however: the painstaking faux-analog sound of *Nation's Pride*, the movie within a movie, for its sound-geek appeal.

Like *Gravity* in 2013, *The Hurt Locker* builds an otherworldly environment in which humans are intoxicated, in part by being deprived of oxygen. You can hear this danger much better than you can see it, and it falls to digital sound editors to exploit its dimensions.

What a great challenge in sound design. As many of us discovered on analog phones in the twentieth century, the various sounds of breath—gasping, sighing, speaking, expiring—may be the first and most consequential sound effect. To capture *in sound* the living-dying body is to express the magic of digitization while gesturing tantalizingly at the loss. This may be how music is successfully reconstituting itself for the digital age.

6

EVEN IF YOU DON'T BELIEVE IN IT

THE TUMBLER

In 2013 a team at Google, Inc., set out to hack death. It was a logical extension of all of Google's ideas, especially its confidence in information and the art of organizing it. The team tabulates the data that the human body throws off—blood sugar, heart rate, hair-follicle circumference—and then aims to prevent the contractions and expansions of heart, brain, and lung from ever going to zero.

This is Calico: a secretive venture, headed by Google's cofounder Larry Page and dedicated to the proposition that mortality is a disease with a cure. To those with operatic and ultramanly visions of the future in the Elon Musk style—Mars, driverless cars—the prospect of being dead before the tech-norapture is profoundly vexing. So why die? Technologists at Calico take a can-do attitude toward their project. They treat aging as a few thousand downer lines of code to circumnavi-gate. Ray Kurzweil, the author of *The Singularity Is Near* and hawker these days of every kind of potion and pill to upgrade "the millennia-old software" that drags down our bodies, figures prominently in the project.

Calico's is an approach to mortality for supermen. These are the types with enough money and muscle to pursue vainglori-ous ends without anyone questioning their ethics or their sanity. Calico men are the latest incarnation of the charming megalo-maniacs who used to freeze their own brain or angle in other Faustian ways to extend terrestrial term limits.

But then I came to believe in technoheaven myself, sort of. I caught a glimpse of it, anyway, when I spilled a tumbler full of water into the keyboard of my laptop. That was the same week that *Proof of Heaven*, a book about a near-death experience by a brain surgeon named Eben Alexander, hit the bestseller list. I had been reading it when I accidentally drowned my keyboard. I was also not believing it.

Nestled among beams of divine light, Alexander's heaven happened to star an amorous peasant in Corot-like colors. This maiden was suspiciously sultry. She was my problem with the whole scheme: she seemed like the decorous doctor's answer to the virgin-dense al-Qaeda paradise. I decided Alexander's heaven was too cartoony and charged with longing, too much as we'd want heaven to be to be how heaven actually is.

The MacBook Pro shorted. There was visible smoke. Familiar self-reprimands reeled through my head: *I have to be more careful. How can I mistreat such expensive things?* Heart pounding, I scooped up the laptop like a bunny to the vet's. My neighbor Sarah came downstairs to watch my children. Off the MacBook and I went in a taxi to Manhattan, to the gleaming multistory Apple Store in Soho.

To get through the Alexander heaven memoir I had suspended judgment about its argument. I had not weighed it as true or false. Instead *Proof of Heaven* hit my brain like fiction, and somewhere comfortably in my limbic system it registered as forged history. At the same time I couldn't help but notice the strenuousness with which Alexander established his bona fides as a skeptic. There was something excessive in it: way too much protest. He was a lifelong man of science, or so he kept saying in a sort of "rationality topos." (The "modesty topos" in literature is a shucksy rhetorical move whereby someone establishes that he's just an average Joe. A "rationality topos" could analogously be a spot where a scientist sets himself up as unfoolable, only to argue for UFOs or paradise.) As a man of science, Dr. Eben Alexander was not prone to unreason in the least—until his near-death experience.

What happened was this: When Alexander slipped into a zero-brain-activity coma, when bugs had eaten his brain to lifelessness, he remained conscious. He met the girl. He saw beautiful lights, and he knew that God was good. He concluded that consciousness survives bodily death. This is a modest argument, in the end—modest as a smiling peasant girl in dark green with peach-colored trim. As I came to believe, Alexander's argument was also, as the logical positivists used to say, unfalsifiable.

My laptop was finished. The Apple Genius (the Romans believed geniuses were liminal figures between gods and humans) didn't mince words: the logic boards and other firmware were scorched. Nothing in the computer could be animated or sparked ever again. My throat ached, the familiar beginning of grief for data and expensive commodities. All 1,624 grams of this Cupertino-designed matter, once so stylish and radiant with meaning as to seem soulful, were now dead as a doornail. But the Genius consoled me with these words: "Your device may have failed. But everything you care about is in the Cloud."

And with that I discovered the apotheosis of data, the instant when pixels and bytes quicken into divinity. Text, images, film, and music had endured the death of the three-dimensional, time-and-space-occupying matter in which they were engraved. The stuff of consciousness—art!—persisted, here among these Geniuses. The truth of the Cloud was at that moment made real to me—to a mind, admittedly, more than usually inspired by the romance of technology.

Consciousness had survived bodily death.

I visited my various Cloud services, and there it all was: pictures of my children auto-saved to Facebook, songs composed

by friends on SoundCloud, letters and poems and the manu-
script for this book on Google Drive.

WILLIAM JAMES'S MOTHER-SEA

Like many Americans, I have a track record with religion that
syncs with the case studies described in *The Varieties of Religious
Experience* by the nineteenth-century American psychologist
William James. That's not because I *derived* my beliefs from that
book, as if from the *Tao* or *Dianetics*. In fact I thoroughly read
James's book only a few years ago. Rather it's because the book—
which tells, in a rollicking way, of all manner of revelations and
religious visions—is purely descriptive. It describes what Americans already do, and in short what many of us do is this: we
believe what works. What works to lift suffering. What works
to consolidate families. What works to secure a class position.
What works to impress our friends. What works to calm our
nerves or build our egos or give us hope.

That kind of nonideological, pragmatic thinking also per-
vades digital existence. Very few of us with dense digital lives
ever fully grasp the concept of the ether, of machine code, of
why or whether computers are an existential threat to humans
or liberal-humanist values. We don't know why sometimes a
hard reboot evaporates a problem. Or why, at the level of silicon
and nickel, one phone is superior to another.

As much as tech consumers and critics drone on about
specs, technology thrives not on empirical facts but on what

neo-Marxists like Walter Benjamin used to call "aura." Mid-century leftists considered this aura—that quality that made a work of art irreplaceable and precious and available only to the elite—undesirable. It was meant to be "withered," in Benjamin's sense, by the industrial technology of reproduction. You'd think digital technology would have finished off what the printing press started, but it has resurrected the aura like nothing else. Witness the rise of homemade objects, as on Etsy. Or think of someone who has mislaid his phone, or wrecked it. He weeps, as for his very soul.

Fortunately the lesson of my water-spoiled computer showed me that the secretions of my mind might not in fact inhere in matter but rather live forever in the Cloud. And the spectral world of the Cloud is not unlike the heaven James describes. I had encountered James's little-known heaven while researching my dissertation, "The Threat of American Life," in 2001. I couldn't forget his description of heaven, or his complaints about it. This particular heaven can be found in a mind-blowing lecture he gave at Harvard in 1897 called "Human Immortality." It was humbling, even embarrassing to talk about immortality at the time, and the title was not James's idea. At the turn of the century the country's ranking biologists and philosophers were almost all "cerebralistic materialists." They believed, as most reasonable people believe today, that life ends when brain activity does. So James faced a tough crowd. But his needle-threading lecture is among his most gorgeous rhetorical productions. He eventually argued that individual brains do not themselves represent consciousness. Rather a single brain is just a filter for something he calls the "mother-sea" of shared, collective consciousness. The filters die when a brain dies. But the mother-sea endures.

Hm. This idea might please a Buddhist, but it was cold comfort to James's Christianish audience. What good American individualist with a sturdy sense of self wants to live forever as a mother-sea blur? I imagine, for example, that Google's Calico developers generally want to live forever as themselves, as Ray, Larry, and Sergey—or even as trimmer, sharper, more euphoric versions of Ray, Larry, and Sergey. James would have bitterly disappointed these men; he had abolished from his eschatology the very creature—the individual—who he was supposed to argue would never die.

But because this is the cunning pragmatist William James, that's where the argument about vaporous cosmological subjects turns political and practical. "Human Immortality" becomes an exhortation to accept a broader vision of humanity and civilization. Around 1900 James's audience was already being forced to accept the fact of mass immigration to the United States. James asked his listeners to shed those elitist selves we define as owners, getters, and trademark holders and recognize instead our part in the vast bank of deathless consciousness. All can be taken in at that bank. Every refugee, every derelict, every pickpocket. No one of us is "indigestible" by heaven, as James puts it, comparing heaven to the day's jammed ports for new Americans. We might call this place heaven or a mother-sea. Or maybe (using today's pious capitalizations) it's the Internet, the Cloud.

In twenty-five years the Internet has doggedly modeled for us a strange but familiar truth: that our lives are both here—in our physical beating hearts; our thick-skinned, small-boned feet; our despised fat; our buzzing brains—and elsewhere, in a fathomless realm channeled through our phones and laptops that we can but dimly intuit.

HOUSTON EXURB

The concept of Cloud computing was minted in November 1996 not in a groovy Los Altos garage or a feverish Harvard dorm room but in one of those Platonic nonplaces: an office park in an exurb of Houston.

In Web 1.0 days a small group of Compaq executives in Texas slipped the phrase into a yawner document, "Internet Solutions Divisions Strategy for Cloud Computing," that predicted, shrewdly, that consumers would one day save data to off-site storage centers maintained by third parties. (The document exists; credit has been claimed.)

It doesn't surprise me that the Cloud was first detected in Texas. Technology is always regionally inflected. Like architecture, it expresses local folklore. The products of Cupertino, California (Apple's Macs and iPhones), express the would-be Buddhist mysticism of the late Steve Jobs. The products of Mountain View, California (Google's Gmail and search), express the Jewish progressivism of Stanfordites Sergey Brin and Larry Page. In this millennium we have become increasingly accustomed to thinking of technology as cradled in Silicon Valley and aesthetically inflected by Buddhism, Judaism, and liberalism. But something born in Houston is bound by the Baptist-derived religiosity of the region—a theology heavily determined by anxiety about sin and death. Rod Canion, the cofounder and first president and CEO of Compaq, also sits on the board of directors of Young Life, an evangelical ministry started in Gainesville, Texas. It's worth noting that the "Statement of Faith" for Young Life says this: "Those who are apart from Christ shall be eternally separated from God's presence, but the redeemed shall live and reign with Him forever." Live and reign on Compaq's Cloud.

A FAILED ATHEIST

In 2014 I had been living with the Internet in one form or another for thirty-five years, from my preteen adventures on Conference XYZ to my work reviewing mobile apps like Square Cash and Device6 for Yahoo! News. The ethereal leanings—leanings, literally, toward the *ether*—of tech developers, evangelists, critics, engineers, executives, and hackers were familiar to me. I saw them in myself. "Electronic circuitry," as Marshall McLuhan called it, was doing something to my nervous system, something not quite empirical and not quite measurable.

I had studied McLuhan's stunty musings since college. I knew him to be a Roman Catholic mystic, whose far-out ideas of media as the "extensions of man" derived from the anthropology of the beaver trade in remotest Canada. How did such far-flung, lonesome, and polyglot traders build a market—a shared space, with transmissible values—if not through the air, through the ether, through strange systems of semiotics that seductively eluded explanation?

Many other writers on media came to the subject through the study of the drumbeats of East Africa. Some came through the study of church bells and prayer calls and how they communicate across vast distances. How are time and space compressed through sound, semaphore, and symbol?

Something else drew me to media studies: the dynamics of proselytizing and propaganda, the selling of ideas. Proselytizing by the left, by the right, by professors of political science, meteorology, and women's studies. Propaganda by Goebbels, by the Vatican, by Zionists, feminists, Wieden + Kennedy. In the 1980s and 1990s you'd have had to cut every single high school and college class not to have been influenced by postwar teachers and their obsession—shaded by a determination not to let the Holocaust happen again—with how minds get made up. My teachers, of various political stripes, in various disciplines, perseverated on tyranny, authority, mind control, brainwashing, advertising, and duty—so I did too.

So how do we make up our minds? Long after my formal studies I found the answer in an unlikely place: a documentary I alluded to earlier, *The Tailenders*. For the film, the director studied a group of evangelical missionaries determined to translate especially persuasive Bible stories into every known language.

The missionaries' smugness and salesmanship tend to irritate other humanitarian workers, who understandably see themselves as more respectful of the people they're tending to. What's more, the film implies, silencing the stomping beats of, say, the Solomon Islands in favor of pallid "Jesus Loves Me" sing-alongs seems just wrong.

More disturbing is the psychological and spiritual danger that many progressives believe is wrought by missionaries, who swipe from indigenous people their happy, peaceful ways and stick them with the greed, selfishness, jealousy, and wrecked natural landscapes known to be the key features of global industrial capitalism. Despite a century of such complaints, however, Protestant missionaries persist. And they're dogged. They dress in uncool hiking clothes and pack uncool backpacks and buses with uncool food and uncool Bibles and venture way the heck into the jungle where they—and this is the subject of *The Tailenders*—learn thorny indigenous languages so they can actually talk with people who have never heard of America, capitalism, jihad, McWorld, or Jesus Christ. Missionaries may be the most parochial and audacious avatars of our modern world.

The focus of *The Tailenders* is the Global Recordings Network, founded in 1939 in Los Angeles by an evangelical named Joy Ridderhof. She wanted to disseminate Bible stories via phonographs and gramophones. In the film still photographs bring to life her adventures among those she aimed to convert; there she crouches, pale and delicate, with various less delicate-looking figures in jungles and on beaches, marveling at a tape recorder. Of the eight thousand languages and dialects believed to exist, Global Recordings has now produced Christian propaganda in more than 5,485 of them. No linguistics department could pull this off.

The idea of releasing disembodied sound on unsuspecting people—like God in the burning bush—clearly fascinates Horne, who conveys an infectious sense of "This blows my mind." The ingenious hand-cranked audio devices, engineered to be usable by people without electricity, are presented with the amazement that only a filmmaker pious about audiovisual technology could convey. "Every physical movement and action reverberates throughout time and space, for good or ill," says the spacey- and sad-sounding narrator, finding an analogy for the way sound echoes. "The ripple on the ocean's surface caused by a gentle breeze and the deeper furrow of a ponderous slave ship are equally indelible." This surprisingly lovely poetry is anchored by down-to-earth reporting in India, Mexico, and the Solomon Islands. At one point a missionary is translating a message about Christian redemption into dialect. A native speaker finds an error; he tells the missionary that the message now says, "We will wash away God's sins." Something needs to change. The voice-over says, "Where Protestant missionaries go, industrial capitalism follows. To convert to evangelicalism is to replace indigenous collectivity with the pursuit of individual economic gain."

And then there's a lament for what's lost. One of the converts says that the new Protestants are shunned by their village; they've forgone the religion of their parents. Only if you've been watching closely will you realize that that lost religion is Roman Catholicism. These congregants have not lost tribal practices; they've just moved on from the last wave of colonial proselytizing.

If you, like me, like to draw out moments of wonder over such matters, marveling at the mysteries of consciousness, shared and separate, and the unlikeliness of death's finality, given the ether, given art, you are a likely candidate for technology theory. Perhaps one day you read about Ray Kurzweil's demented confidence in the Singularity and then his further demented confidence that death might be cheated. (At the end of 2013 Kurzweil cryptically told a reporter of his prospects for eternal life, "I think I have a good chance—I would put it at 80 percent—of getting to the point where it becomes indefinite, because you'll be adding more time than is going by to your remaining life expectancy.") Maybe you were drawn to the sci-fi trickster Arthur C. Clarke, who cowrote *2001: A Space Odyssey* and posited three laws of—of *something*, culminating in "Any sufficiently advanced technology is indistinguishable from magic." I know I was.

At the same time, my own religious life had taken some unlikely turns. The daughter of a Methodist-cum-Episcopalian mother and a Roman Catholic father authorized to take Communion at the Anglican Church (he insisted he'd gotten "special dispensation" to take the sacrament with his wife), I was baptized at the age of seven as an Episcopalian. In preparation for my confirmation as a teenager I served as an altar girl—an acolyte, in the Church's jargon—and took theology classes with our priest. In my junior year in high school I started to seek intoxication and romance at Dartmouth College. I got decent grades in biology, chemistry, physics, English, and social studies. Acting out in ways that involved college fraternities, cocaine, and the town sheriff seemed somehow to go hand-in-hand with working hard in school in hopes of attending a drinking school with an elitist sheen. I understand it's not the same for today's ambitious high schoolers; now to get in to a good school, you have to be good on all fronts, including the behavioral ones.

In any case all of the scientific experiments we did at Hanover High School imparted a distinct incandescence to components of experience on earth I might have overlooked: cells, elements, trajectories. I especially liked learning how a tube television works. The diagrams deepened my awe, the more for their insufficiency to explain how moved I could be at the sight of the old lipstick reds and kiddie-pool blues of analog TV sets.

The material explanations of life made modest sense, as far as they went, of the firmware, the chemicals, and the meat in a mostly plodding and occasionally brilliant way. Art, pop culture, religion, and especially literature illuminated the rest. I disliked only the kind of science that seemed to be formed in angry reaction to the more charismatic claims of literature. It came off like an engineer square trying to win you over by telling you your poet bass-player crush is a flake and only he's going to pick you up on time and buy you a McMansion.

Religion continued to fascinate me. But when I got to the University of Virginia, Christianity turned both frightening and boring. There were too few Roman Catholics to keep things rigorous or emotionally exciting, and almost no Jesuits or Jews, whom I'd come to believe held the keys to actual mental pleasures. For the first time I heard people identify as simply nonsectarian "Christians," of what I didn't know then was the Compaq–George W. Bush variety. At length I came to understand this meant evangelicals or Baptists. My first-year roommate, an extremely gifted triple-varsity athlete (volleyball, basketball, and softball), was a member of the Fellowship of Christian Athletes from her first year on. She taped a poster to our wall about scoring goals for God.

As an intellectually restless Northerner both depressed by and envious of the (apparently) contented minds around me, I realized that the most sensible move for me was to embrace atheism. I had sworn off computers, after all, which I loved, because I feared that otaku passion would cost me popularity with the girls and guys I admired. I reasoned that I might yet muster the same sangfroid and swear off God.

Lacking my own Mac, I brought a cumbersome Apple 2e to college, for word processing, but I also laid out on our dorm-room floor a purplish Muslim prayer rug I'd haggled for in Marrakech and taped up prints of ashcan school paintings of New York City, which I'd hardly ever visited, as well as "tapestries," the thin cotton Indian-printed blankets that boarding-school kids in the North used to impart an opium-den ambience to their rooms. I wore amulets and oversized cotton tops. Still aglow from the summer's trip to Morocco with my family, my goal was to be earthy, sexy in a hippie-girl way. My plan was also to never, ever mention computers, the anticharismatic Soviet or IBM nerd machines that were at steep odds with this pose.

Atheism didn't take. I studied philosophy. In seminars with names like Shakespeare and Evil and Faith and Doubt, I listened intently to buzz-cut, rangy guys and their punk insistence on God's nonexistence. Their arguments produced in me a metallic, sad feeling. Brad Braxton, an unforgettable Eagle Scout, black, who wanted to be a preacher, was also in Faith and Doubt; he ended up studying at Oxford on a Rhodes Scholarship, and later running the Religion in the Public Sphere program at the Ford Foundation. Instead of making comments in the seminar, Brad preached gallantly in the King mode. A devout sweetheart named Anne was his witness.

One atheist in the class had the air of a vandal. He smoked like I did. One night he punched a mirror and showed up with a bloody hand; I was led to believe the injury was meant to show me something. The other atheists he hung out with were healthy, close-cropped types in button-downs. They liked to scoff at believers, though Brad had them in a bind, since he was joyful, brilliant, and nearly impossible to dislike without seeming racist.

The metal-taste misery surfaced in me every time the atheists spoke with malice about the idiocy of theism. A better mind than mine might have gone *toward* my sorrow—truth at all costs—but I sensed that the atheist teenagers were not better than I in the sense that they had heroically endured the metallic sadness that came from denying God. Instead their atheism, and especially their contempt for believers, produced pleasure for them. It was a Fight Club kind of thing, and because I was not immune to the charms of punk and smoking and hating I could see how it could be fun. But I couldn't brook the sadness, or the nausea.

Week after week in the seminar we read Kierkegaard and Bertrand Russell and A. J. Ayer and danced around Brad's blackness and Anne's prettiness and the atheist guys' alienation and spite. Eventually I sensed that not one of us had looked deeply into the universe and seen—whether with a telescope or the soul—the truth of God's existence or nonexistence. Instead we were assuming a set of seminar poses meant to impress or estrange one another.

Also, I missed computers. Though artificial intelligence was the specialty of at least one member of the Philosophy Department, we never talked robotics or computers. I read about those things on the sly. I spent time with my friend Peter Cousins, who had been a prodigy programmer with a Commodore 64 and then had a real job thwarting private calls by employees at MCI (at the age of thirteen!). Like me, Peter was studying the humanities in the explicit hope of becoming more human, saner, and more attractive to his peers. Now a prosperous and insightful tech executive, in those days he had already read Hubert L. Dreyfus's *What Computers Can't Do*. I remember our long walks and confessing to each other our pre-Web fascination with digital technology as if confiding that we were both anorexic.

(Later, in 1993, Peter predicted that in short order—ten years, tops—we'd spend 80 percent of our waking hours "in cyberspace." When he told me this, I think I saw stars, as if I'd been clubbed. I remember right where I was, as people remember where they are whenever they get news that blows their circuits. I suspected he was right. He was.)

Not until my senior year, in 1991, was I invited to use a computer program as part of a class. The class was Formal Logic, and it was required for my philosophy major. The program let you practice logical proofs, and it was so beautiful and fast I almost ached to use it. The green backlit symbols on the screen were just like the ones on my old Zenith Z-19, and the dark yonder evoked a vast and mysterious world. All possibility.

To watch the program work and work and work again; to find without emotion the inconsistencies in my twenty-two-year-old applications of ancient principles of modus ponens and modus tollens to Ps and Qs; to see the *formality* of it, the perfection of the code, was—I'm channeling some college-age pretensions here—like witnessing the wildly symmetrical Saharan dunes from the vantage of the High Atlas Mountains in North Africa. Twee or crazy as it may sound, a time or two I took a tablet of ecstasy and snuck into the computer room, fired up one of the massive terminals, and signed in to spend the night in a solo orgy of logical proofs.

My boyfriend was Jewish, or something like Jewish. His

mother, a Methodist from Appalachia like my own, had converted to Judaism when she married his father. In time she'd found the conversion stifling and acted out, spying on her many children, and marching them first into Jews for Jesus and then into a Christian cult. Late in their lives she baptized them in Israel. My boyfriend disliked the strain religion had put on his family, but atheism held no comfort for him. His Jewish past, forever in tension with his mother's cultish practices, seemed to give him a seriousness and even a piety I hugely admired.

Seriousness to me was fast becoming coded as Jewish. In the aftermath of the Holocaust, many academic Jews had come to Dartmouth; they sat at my parents' dinner table and told stories of Eastern Europe and joked in a warm and cerebral way. I thought Ray Sobel, a psychiatrist, and Steven Sher, a Hungarian musicologist, were the wisest and funniest people I had ever met. They were not presentational and alcoholic like many of the gentile academics who came to our house for sherry and adultery. They were radical lefties, *Nation* subscribers—and atheists. I heard them speak of atheism and affectionately deride Americans, including my good-natured mother, as churchy, but they seemed holy in their own way. Injured and reverent and also joyful. I didn't take their denial of God as evidence that they weren't religious. Perhaps they worshiped art or the human spirit. (I thought that was a bullshit workaround, but I didn't begrudge these loud-laughing Europeans anything.) Ultimately the Jewish lefties were not iconoclasts, and there was nothing dry and empirical in their speech. They hated their dopey students and the idiot GOP in a way I liked, even if, in my ever-present desire to be a regular American, I didn't allow myself that same hate.

WHEREOF WE CANNOT SPEAK

As I became more academic myself, and more philosophical, I was drawn to the Jewish intellectual tradition. Leaving off Freud and Marx as too popular, I selected Ludwig Wittgenstein as my first philosophical hero. A professor warned me against becoming "sycophantic to Wittgenstein" (I had to look it up), but I became sycophantic nonetheless. Wittgenstein had known Hitler. He wrote alongside Bertrand Russell and the other English (atheist) philosophers, even Ayer and the other supremely silly logical positivists, but was not *of* those philosophers in the least.

In those days Wittgenstein was how a philosophy major like me, who had urgent late-adolescent questions about existence and ethics that were not being addressed by British logical positivist jive, allowed herself to embrace a quotient of mysticism again. A close reading of Wittgenstein's *Blue and Brown Books* reduced thinking to "talking and suppressing the noise," as my lecherous but shrewd professor put it. Wittgenstein also generously let language off the hook for corresponding to reality.

We were not apprehending reality when we "did philosophy," or even when we reflected deeply on important matters. Wittgenstein exposed philosophy itself as a "language game"—a phrase that touched every sweet spot for me and produced a thrill akin to the one I felt in the computer room with the formal logic program. Language. Games. The two highest treats of existence. I needed nothing else.

Not all philosophy majors at the University of Virginia were so excited by Wittgenstein. In those days the department employed a clique of Oxford men who were analytic philosophers, as opposed to flaky, far-out continental philosophers—the French deconstruction crowd whose tortured English, snooty pose, and endless obscurantism was easy to parody. Not only did these men appear to care about weird questions—"Why does a penny sometimes look like an ellipse?"—but they hotly debated the canon of possible answers, especially the hypothesis that you didn't actually see a penny when you saw a penny; instead you saw a little floating thing called a sense datum. These belligerent dons managed to keep their curiosity piqued on this kind of esoterica. For a time I believed that this was what a great philosopher must do.

But then Wittgenstein became all I needed to part from the atheist chumps with the monologues about sense data and mind-body nonproblems. "Whereof one cannot speak, thereof one must be silent," wrote Wittgenstein early in his career. What a philosophy! Shut up about God or "God" or no-god, I thought.

I rushed headlong into the sunlight of the Department of the Humanities, a department of one: the great pragmatist philosopher Richard Rorty, whose survey course I took while I was also reading *Hamlet* for a seminar in the English Department. It had occurred to me that Hamlet's problem, like mine, was that he couldn't stop talking: he was too interested in every problem, from deception to romance to suicide, and his soliloquies were never ending. Professor Rorty helped quiet me—and I will never not be grateful for that. For one, he was indifferent to the penny as ellipse. He believed that analytic philosophy had lost its way; he flaunted his obliviousness to its arcana. He also thought the hocus-pocus of deconstruction was a little much. Instead he introduced our university to his solo project, the Department of the Humanities. From there he showed off his shrug.

"I'm sorry," an undergraduate stammered in the first discussion section to his philosophy survey course. "Is it pronounced 'Berkeley' or 'Barkley'?"

Rorty shrugged, his chin doubling. "You can say 'Berkeley' or 'Barkley,' but I think Berkeley said 'Barkley.'"

With shrugs, admissions of ignorance, and bland incuriosity, Rorty encouraged his students just to *drop it* already. So many para-intellectual anxieties are a waste of time, he let us know, to say nothing of the genuinely intellectual pursuits that represent cosmic and often lifelong wastes of time. Among these, he made clear, were the so-called cogito, capitalism's base and superstructure, the mind-body problem, the Gaze of the Other, and being subversive in a world of hegemony. Make your private life beautiful and your public life humane, he taught us. This coherent approach to—to—life itself Rorty rigorously justified in his many books. I believe he argued it most simply and persuasively in the opening chapters of *Contingency, Irony, and Solidarity*, as well as in his personal essay "Trotsky and the Wild Orchids" and his lecture on Irving Howe.

Rorty's exhilarating survey course—From Hegel to Derrida, if I remember right—really did place the period on a gasping run-on sentence I'd been writing and speaking since adolescence. Now I could shrug, and live. I quit philosophy right then, moved into literary criticism, and aimed to talk less.

By 1990, when I found them, the deconstructionist literary critics, including especially Jacques Derrida and Paul de Man, took for granted the noncorrespondence of language to reality. Instead of panicking they seemed to enjoy when the metaphorical content of sentences undermined their empirical, upright meaning. Now when the atheists I knew spoke of, say, "the girlish hope for a protective God to buffer death for us in what is in fact a bleak unprotected universe," I heard them not describing the essential cold facts of the cosmos with 20/20 vision, as if they were using equations, but talking about gender and family and fear and rivalry and sexuality and paternalism and maybe some other stuff like facticity and the consequences of buffering. I liked hearing the language, in its deep and relentless metaphoricity, take on a mind of its own and subvert the best-laid plans of writers and speakers to talk empirically and with authority. That was the game part.

I kept Wittgenstein close: "Whereof we cannot speak, thereof we must be silent." This was part of an early book whose arguments he moved on from, but the heartbreakingly perfect tautology has grounded my thinking since I encountered it. I had had it with debating God's existence or the primacy of meta-phor, which now seemed to me self-evident. I had lost my taste for debating society engagement, wherein I would point out all the garish poetry mischievously clouding the philosophical statements that purported to be transparent descriptions of the thing-in-itself (the German *Ding an sich*).

Whenever I tried to tell philosophers that they were more like poets than scientists, I would be accused, over and over, of embracing stupidity and willing bodily injury by refusing to ac-cept the reality of, say, a cliff or gravity. "How could I refuse the conceit of gravity," I would ask. "Conceits?!?!" my philosophy professors would shout—often actually shout; using literary lan-guage in philosophy class triggered red-faced rage. They would then go to great lengths to humiliate me by asking if a newspa-per on the desk was really there, or not really there, whichever would make a greater fool of me.

Though many students seemed to find amusement in them, these set pieces were stalemates. No one, least of all me, wanted to be convinced of anything. Philosophy was behind me; I re-solved to do graduate work in English. No more squabbling about the reality of God or the reality of reality or sense data. Whereof I couldn't speak, thereof I would be silent.

PAVEMENT, LIZ PHAIR

"Anything from the sound of a word through the color of a leaf to the feel of a piece of skin can . . . serve to dramatize and crystallize a human being's sense of self-identity," wrote Rorty in 1989. Those words themselves seemed to dramatize and crystallize something for me, as did Rorty's elegant bifurcation of one's ideological life from one's aesthetic life. I might find peace and self-identity—a soul made palpable—in the touch of a friend's hand rather than in an argument about consciousness made by Daniel Dennett (Rorty's great frenemy). This of course made intuitive sense, but until I encountered the case for it in *Contingency, Irony, and Solidarity* I wouldn't allow myself to believe it. I wouldn't have dared.

Sounds, leaves, and certainly skin were girl stuff. Nothing so fleeting and irregular and sensory should ground something so fearsome as an *identity*, or should it? Rorty appeared on the cover of his book surrounded by flowers, yet he was an august philosopher. I liked his view; it washed away the metal melancholy of the punk fake-empiricist gang, not to mention the popular idea that our identity is a function of power relations around race, class, and gender. And so I reveled in it.

When I thought about the rapture implicit in accepting the contingency of identity—its happenstance, its randomness— I liked thinking about Erich Auerbach too, who composed his ingenious work *Mimesis* in Turkey in exile from the Nazis with only a Bible at hand. I also thought of Jack Henry Abbott, the prisoner championed by Norman Mailer, who found himself visually deprived in a strip cell, "so fucked up even the sight of a piece of colored cloth moved me to euphoria."

Fragments of Pavement and Liz Phair songs, the taste of

bourbon, the cheeks of a face I loved did dramatize my identity, gave rise to my best ideas, and could readily move me to euphoria. I was learning to pursue thoughts, beliefs, and things that made me productive and happy, not miserable. That was the newfound American pragmatist in me. I also learned to detach, with something like love, from Western scientistic philosophy. This was a great moment in my growth as a person and, in a minor way, as an ex-philosopher.

LOW-BROW

If a colored cloth or the hiss of a cassette tape could move me to euphoria or crystallize my identity, then certainly the O.J. trial could. And Anna Nicole Smith and *Tootsie* and my first five or six emails.

It was an article of faith with me, as it was with my professors in the 1990s, that any cultural artifact (and even some natural ones), no matter its origin or class valence, would respond to critical attention. Maybe I was cloistered, but it became hard to remember a time when Wordsworth was thought to express a poetics and Victorian print advertising was not. Thus I arrived at Harvard to start graduate work in English and American literature and language well aware that we'd be studying Bette Midler, Anita Hill, and Marilyn Monroe along with Milton, Chaucer, Spenser, and Eliot. Any of it, after all, could move us to euphoria or dramatize that elusive self-identity.

Graduate school was filled with many warlocky professors and some wonderful maternal ones. Some of them could persuade you with their comic-villain furtiveness or displays of pity and concern that PhD work did not involve reading chiefly funny, fruity poetry about birds and flutes and fairies; rather it was a gothic and dangerous game, as self-serious as the RAND Corporation. One false move—a wrong inflection, a misplaced reference—could cost you everything and fast-track you for suicide.

The faculty, the low-ceilinged rooms of Warren House, and much of the theory on the syllabus scared the hell out of me. A seminar topic one day was "the erotics of pedagogy"; around the table we were expected to examine our sexual fantasies about our professor. Scared people are scary, and so my colleagues and I, in our early and midtwenties, went ahead and terrified each other. You could get your head bitten off in a seminar, or left out of a party, as I was, near Valentine's Day, when I went to the graduate school lounge and saw pink invitations to something in every mailbox but mine. I felt overstimulated and excited at Harvard but also nervous, visibly trembling. About what, I couldn't say. Social life was brutal. The personal was political. Every interaction seemed to flicker and shape-shift, uneasily, like encounters in Henry James's novels. Told that I'd been let in as a U.Va. graduate in the name of bringing to Harvard "intellectual diversity" from the hinterlands, I lost faith in the combination of pragmatism and poststructuralism that had framed my thinking in Charlottesville. Rorty seemed parochial; Walter Benjamin and Theodor Adorno were the thing, but I didn't have room for another personal intellectual revolution. Much of what

I concluded in my long years in the Harvard English Department was that whatever I thought, it was wrong. Fortunately I also read and read and read.

Much, much later that anticonfidence—faith in my wrongness—became a gift I tapped into. In 2011 it saved me to realize my "best thinking" had landed me in dire straits in my health and my personal life, and thus I would need to cede the mental driver's seat. At the time, though, I wanted only to shake off my conviction in my wrongness—drive it out with a pose of righteousness or drown it out with distractions and red wine.

Improbably enough, one of the times the deep self-doubt did lift was across the campus from Warren House, at the Harvard Law School. A friend had introduced me to a 2L named Jonathan Zittrain. Jonathan and I struck up a friendship at first based on a secret I was keeping: I had recently signed up for CompuServe email. So had he.

An unforgettable string—73773.143@compuserve.com—was my first email address, and I was sure I'd keep it forever. It was 1993, and not many friends had email, but those who did got many, many email messages from me. Stretching a long telephone cable across the floor of my apartment, connecting my Powerbook to the wall, and waiting for the squeal-and-crash of information fired my imagination. Still I told myself I wasn't going to lose my head on the computer this time, as I had in childhood and tweenhood with Conference XYZ. I wasn't going to spend too much time at it or get into masquerade and mind games.

But Jonathan was a real-life friend, and he too used CompuServe. He was a systems operator for CompuServe with a State Department passport, though he wasn't yet twenty-five. He sold a big hank of AOL stock short, he was that persuaded that CompuServe would win the day. His short sell was ill-advised, but he took the loss just fine.

Jonathan invited me to what he called the Cyberlaw Seminar at Harvard. Led by Charlie Nesson, a dashing, silver-haired professor in a black turtleneck, the Cyberlaw Seminar was all improv. Boomer cronies of Nesson like far-out Grateful Dead lyricist John Perry Barlow, who had founded the Electronic Frontier Foundation, came in. Sometimes there were U.S. Department of Justice figures, like Jamie Gorelick, or others from the government. They rapped about what governance in "cyberspace"—where many of us had never been—might look like. "In cyberspace the First Amendment is just a local ordinance," was one dire saying, which rightly predicted wide-scale Internet censorship in China and elsewhere. Another was "Reading on the Internet is like drinking from a fire hose."

I was enchanted. I audited the course and even chipped in thoughts, mostly in the after-seminar dinners, about "the Internet," which I now imagined as a combination of IBM-style data storage, the cartoony CompuServe homepage and email, and chiefly the rich, odd yonder I had seen and sensed in my days of role-playing on the Dartmouth Time-Sharing System. *Cyberspace* seemed like a good word for it. It wasn't written over with ads and graphics and signposts like the World Wide Web of today. It was vast and dark, cold, and describable only in a language so cryptic it was *called* code.

Here we were at grand old Harvard in earnest, reflective steward mode. The Cyberlaw Seminar might have been convened in 1636, the year of Harvard's founding, by aspiring Unitarian and Congregationalist clergy bent on fair governance among the Indians in the Massachusetts Bay Colony. How were *we*, the clerisy, going to scrupulously govern cyberspace—a place that's not a place, a people that can't be located or counted, a weapon posing as code, documents, and speech acts that exist only in thin air? The talk was speculative, outrageous, utopianist, and dystopianist by turns, and above all playful. In charge were a group of left-wing 1960s futurists who loved to trip out on the question of what would be demanded of minds in the future.

This rapturous theorizing stayed with me, while back in my own department the gender analysis of the O. J. Simpson trial and the outrage at Camille Paglia's lusty showboating and Paul de Man's early Naziesque writings—this too was called "theory"—reminded me of the penny-ellipse problem. A mind-waster, a time-waster, and a life-waster. Yes, the work of Helen Vendler, Philip Fisher, Henry Louis Gates Jr., Elaine Scarry, D. A. Miller, and Marc Shell (who became my dissertation adviser) was brilliant and exciting, but in person they often seemed fretful and preoccupied. I decided to take a year off from graduate school and move to New York City.

That's when I decided to become a journalist. I slid into it, working first as a fact-checker at the *New Yorker*, then returned to Harvard to serve as a teaching fellow, start my dissertation, and write for the *Boston Phoenix*, which, like most alt-weeklies, is now deceased. When I moved back to New York with the intention of using Columbia's library for dissertation research and work, I did fact-checking for various writers, and finally and most instructively for Michael Eisner, then the CEO of the Walt Disney Corporation. That led to work at Tina Brown's *Talk*, first as a fact-checker, then as an editor, which in turn led to more editing work, until finally Jacob Weisberg, the editor of *Slate*, proposed I write about television for his online magazine. I loved it. You could write about almost anything under the pretext of writing about television, and I went on to review TV for the *New York Times* for four years, until the Internet mugged me again.

This time it was online video that drew me in—the vast video rain forest that is YouTube. It looked like television to me, and I wanted to write about it, first in a weekly column at the *Times Magazine* and then at Yahoo! News—and write, in turn, about all of digital culture: the infinitely complex music, graphics, photography, film, and literature of the Internet that now form the sections of this book.

First I had to undergo one more giant psychic shift. While at Yahoo! my head was full of the jargon and metaphors of the vigorously applied Internet—search, mobile, social, storage, "dogfooding," "text assets." These our glamorous leader, Marissa Mayer, herself a hacker, used with ease and never translated.

Listening to Mayer evangelize I decided to write a column on why technologists and scientists are different.

For years *technology* had seemed to be the masculine form of the word *culture*. If you wanted to sell men on a culture story, you did well to frame it as a tech story—a story about the plumbing or stockprice of Netflix rather than a story about the pixels that constitute *Bloodline*. Technology is built stuff that aims to be elegant and engaging. Apps are founded on science in the same sense that a watercolor is founded on science, where the chemistry of pigments and the physics of brushstrokes are the science. But the resulting painting, if successful, hints at transcendence or at least luminous silence, something whereof we cannot speak.

The same is true with good technology. Often with digital technology what we sense behind it is a vision of the Internet itself—something so abstract and powerful that we glimpse it through technology the way, Auerbach wrote, we see the face of God in the interstices and lacunae of the Torah.

One hot day in July, I was sitting in my purple cubicle in the Midtown office of Yahoo! News, getting ready to write a fairly light piece in hopes of starting a conversation on something that wasn't sex, internecine tech-world battles, or media gibberish like "mommy wars." I intended to briefly chronicle my development as a tech enthusiast and a nonscientist. In general interest media (as opposed to real scientific journals), *science* seemed to be a byword mostly for cute animal pics, images of space, and health, a topic that pushed on grateful readers both newly manly dieting tips and hypochondriacal visions of our ailing and mistreated American bodies, laced with half-truths from Big Pharma. Social science, I believed, had been hopelessly corrupted by misapplications of scientific method and impressionistic technologies like fMRIs. This had happened most egregiously in the once superstar field of evolutionary psychology and again in the speculative realm of neuroscience.

By then I had comfortably surrounded myself with traditional religionists at my son's Jewish and my daughter's Christian schools. These were doctors, lawyers, writers, and business types who felt true to the religion of their childhood: midcentury American Judaism (often Orthodox in flavor, but they reset it to Conservative) or midcentury American Episcopalianism (what I had grown up with). The father of my children called this crowd the Judeo-Episcopate and claimed with confidence that they casually dominated Harvard, the *New York Times*, and the entire Eastern Seaboard Establishment. I might myself have qualified for the Judeo-Episcopate, but I was way too earnestly theological to be part of that casual, preppie, masters-of-the-universe crowd. I did, however, convert to Judaism to marry my now ex-husband, spent ten years as an observant Jew, and then converted back to Episcopalianism after we divorced.

I was also surrounded by Silicon Valley types who meditate regularly, have spiritual coaches, and listen to Deepak Chopra tapes. Rounding out my group of friends were people in recovery from addiction, sincere seekers for whom prayer was as natural and necessary as physical exercise. Truthfully it had been a long time since I'd been around staunch atheists, except a be-loved high school friend, who lives in New Hampshire and has a bumper sticker that says, "Nothing fails like prayer."

So in July 2013 I wrote "Why I Am a Creationist," using an old word and borrowing the title structure from "Why I Am Not a Christian" by Bertrand Russell. I was trying to be playful and bold. Evidently boldness works better with a negative or if you're Bertrand Russell. The essay was a disaster. Reddit, the giant and tenacious bulletin-board site started by former room-mates at U.Va., led with the story of my heresy, and my week then went off the rails. I entered the Twitter coliseum, where I was roundly mocked. No fewer than four media-gossip sites—Gawker et al.—declared that I was idiotic, dangerous, and fin-ished as a journalist. The storm of invective made me miserable: the angry mail, the concerned mail (the feminist writer Katha Pollitt, a friend, worried for my sanity), the disappointed mail, the this-friendship-is-over mail from the old high school pal. I'd pray for her to change her mind, but nothing fails like prayer. I was hurt and embarrassed, but as the criticism mounted my certainty did too: it had been worth writing. Newspapers in En-gland weighed in: I was maybe not entirely a scoundrel. Finally there was "Jesus was hated too" tweets and Facebook mail invit-ing me to the Creationism Museum in Glen Rose, Texas. That was bracing. *With friends like these.* But I guess I had finally con-nected with readers outside my usual set.

I agreed to appear on *Q*, a Canadian radio program, and at last was able to say what I meant in a more measured tone and with more humility. I told the host that, where physicists typically see a void (before the Big Bang and as the spark that turns matter to mind), I habitually see God. I emphasized habit: I didn't believe my brain was adequate to the task of perceiving the world as it is, so I believed what worked for me. That was American pragmatism. That was Rorty coming in, and William James.

The appearance won a favorable response, but I was tired after the bruising showdown in media and social media. I'd been cast as William Jennings Bryan in a production of *Inherit the Wind* for Twitter. Didn't Bryan die of a heart attack at the end of the film? I knew enough to step out of the hayseed role before the heart attack. I swore off debating *again*. We learn the same lesson over and over till it sticks.

TWITTER LAUREATE

It had been an eventful summer, beginning with my recognition that I'd found a figure for immortality in the Cloud and ending with my statement of faith on Yahoo! News, the wave of antagonism to it, the wave of support for it, and my resolution to shut up on the subject. By chance my religious life and my technological life had both turned mystical and become one.

In 2014 David Thom, the chaplain at MIT and coordinator of the Cambridge Roundtable on Science and Religion, invited me to a roundtable with the philosopher Rebecca Goldstein, who had just published an ingenious book about the persistence of

Platonism called *Plato at the Googleplex*. At the event I was seated around a dinner table with five others who quickly introduced themselves as an atheist physicist at Harvard, a Christian chemist at Harvard, an agnostic engineer at MIT, a trickster mathematician at Harvard, and the Harvard chaplain, presumably a believer. I described myself as someone who believes what works and who had found that a minimalist theism serves her best.

The mathematician picked this up and peppered me with questions: Was I capable of believing things to make myself feel better—without actually thinking them *true*? I stumbled over an answer, but I told him sometimes I did one, sometimes the other, and sometimes both. He brightened. He explained that, like most mathematicians, he was a Platonist, believing that numbers and equations are not figments of the mind or a symbolic game but are "out there," in the natural order. Before Abraham was, they are.

He then explained that he had two ways of solving an equation. Mostly he did so hastily, took shortcuts, made ad hoc moves, and arrived at a finish in a serviceable but not especially sterling way. But on other occasions, when he had time and inclination, he lost himself. Doing math put him in concert with natural laws, in the pocket, as he felt his way through the mind of the universe. Landing on a solution when he worked that way felt like landing a perfect major chord—nothing slipshod about it.

"That must be how it is when you believe in God," he said.

I agreed.

And then peace really came. From the mathematician, and

then in the form of a surprising tweet by Frank Wilczek, a physicist at MIT. During the public back-and-forth over the creationism column, a friend (college buddy Peter Cousins, in fact) had alerted me to the work of Wilczek, who had written equations showing—and here is a summary that is probably violently untrue to the subtlety of his work—that it takes slightly *less* energy to have everything in the universe (you, the trees, opera, Google Hangouts, etc.) than to have nothing. The movement from nothing to something therefore is entropic. For that extraordinary contribution—and really for the equations that gave the idea mathematical form—he won the Nobel Prize for Physics in 2004.

I read Wilczek's great book on the subject, *The Lightness of Being: Math, Ether, and the Unification of Forces*. Reflexively I also looked for Wilczek on Twitter.

There he was: Frank Wilczek @FrankWilczek.

And a tweet: "They say it works even if you don't believe in it."

With those words Wilczek had posted a link and tagged it #amazinggrace.

I sleuthed and read back-tweets. Evidently Wilczek was quoting Niels Bohr, as he had done in an earlier tweet: "Apropos of everything, Bohr's horseshoe: NN: Surely you dont believe that nonsense. Bohr: They say it works even if you don't believe in it."

As I soon found out, that exchange ends Werner Heisenberg's essay "Science and Religion." It's what Bohr said to a neighbor who asked why he was nailing a horseshoe "for luck" to the door of his stables. I was grateful to track down the Bohr reference. And the sentiment "It works" also invokes William James's nonjudgmental chronicle of spiritual encounters that "work" and demonstrably change lives. Wilczek, I learned, also tweets about James. (Let it never be said that Twitter is only for Kardashian gossip.)

This time Wilczek had a special context for "They say it works even if you don't believe in it." He was introducing a link to a video of a Wintley Phipps performance at Carnegie Hall. In the video Phipps, America's marquee gospel performer, shows how a pentatonic scale—the "slave scale" of just the black notes—uniquely expresses "power and pathos." He groans and sighs through a wordless version of "Amazing Grace" intended to reveal the melody's rough beauty and its tension with its smooth, Sunday school lyrics, which were written by a reformed slave-ship captain.

Through all this: "They say it works even if you don't believe in it." Or so tweets the laureate Frank Wilczek, whose Wikipedia entry lists him as an agnostic. I couldn't fight off the tears.

I read more Wilczek, including his criticism of the philosopher Thomas Nagel, whose hedged atheism Wilczek faulted for "advancing an idealist position." At this revelation of Wilczek's allergy to philosophical idealism, let alone frank theism, I had to tweet at the physicist—not for a second forgetting what a privilege it is to be able to address an academic hero this way (and what a useful brake on vanity it is to have to keep one's queries to

140 characters). Thus I asked @FrankWilczek, "I habitually see God in narrative voids—lacunae—'whereof we cannot speak.' Is this advancing an idealist position?"

Tagging me @page88, Wilczek shot back, "It's a gift if you can do it. I just see narrative voids."

It's a gift if you can do it. They say it works even if you don't believe in it. Whereof we cannot speak, thereof we must be silent.

I thanked Professor Wilczek and left it at that.

Triggering the rage of atheists and science writers—the same well-meaning empiricists who hazed me in college—led me to a Nobel-winning physicist, who in turn led me back to the tradition of Rorty, James, and Wittgenstein, as well as my own pragmatic technospirituality, which is made of intimate and comforting language games that somehow work for me. To celebrate I downloaded the latest book of essays by the great novelist Marilynne Robinson, whose book *Gilead* was the first I read on the Kindle. Many of the essays in the new book concern Robinson's Christian faith. In about a minute her prose suffused my Kindle Cloud Reader.

The Internet suggests immortality—comes just shy of promising it—with its magic. With its readability and persistence of data. With its suggestion of universal connectedness. With its disembodied images and sounds. And then, just as suddenly, it stirs grief: the deep feeling that digitization has cost us something very profound. That connectedness is illusory; that we're all more alone than ever. That our shortcomings and our suffering are all the more painful because they're built in the mirror of a fathomless and godlike medium that doesn't suffer, that knows everything, that shows us no mercy or compassion. In those moments death shows through in the regular gaps in Internet service, and it's more harrowing than ever.

Magic and loss, however, have always coexisted in aesthetic experience. Maybe they *are* aesthetic experience. And so I turn back to art barely worthy of being called art: YouTube. That gospel performance by Wintley Phipps really does blow the heart open. It works even if you don't believe in it. Each time I view it I can feel the mysterious and maddening Internet, which I have loved my whole life, throw off—like a sparking logic board, like the glass threads of fiber optics, like a meteor shower of pixels— a measure of amazing grace.

ACKNOWLEDGMENTS

I am deeply grateful for the practical assistance, mind-altering insight, and profound kindness of Mike Albo, Dominic Anfuso, Asli U. Bali, Ginia Bellafante, Tessa Blake, Johanna Block, Sophie Block, John Brockman, Max Brockman, Nell Casey, Caucus, Maggie Chandler, Heidi Chavac, Soraya Darabi, Amar Deol, Susan Dominus, Arin Epstein, Paul Ford, Casey Greenfield, The Grat List, Andrew Heffernan, James Heffernan, Nancy Coffey Heffernan, Megan Hellerer, Allison Johnson, Samantha Knowlton, Scott Labby, David Lavin, Steven Levy, Angie Lieber, Ben Loehnen, Alexis Lotzko, Gerry Marzorati, Sarah Mills, Bob Nastanovich, Alana Newhouse, Heather Quinlan, Hilary Redmon, Heidi Rose Robbins, Michael Robbins, Meredith Rollins, Lucinda Rosenfeld, Alexandra Samuel, Alessandra Stanley, Jamie Ryerson, Shachar Gillat Scott, Geoff Shandler, Jake Silverstein, Brianna Snyder, Susan Stava, Jackie Stone, Chris Suellentrop, Lorraine Tobias, Dana Trocker, Andrew

Unger, Bill Wasik, Amy Witting, Asia Wright, Charles Yao, and Forest Young.

My gratitude for all things goes to my beloved partner, the ingenious and, above all, good Jamie Block. Finally, this book is dedicated to my magic children, Ben and Susannah, who have my heart.

INDEX

INDEX

INDEX

IBM, 38, 42, 47, 218
Iceland, 124, 125
icons, 24, 39, 42
  design of, 43–44
  Google, 44
  Macintosh, 43, 44
  Microsoft, 44
illusionism, 155
images, 107, 111–31
  communication with, 118–19
  composite, 125
  digital, 123–27
  double, 125
  "tags" connected with, 147
  3D, 5, 155–57, 163
  see also art; painting; photography;
    television; videos
imagination, 38, 230
iMessage, 43
"I'm Not a Photographer" (Merkley),
  125–26
India, 214
Inherit the Wind (Lawrence and Lee),
  237
Instagram, 5, 7, 12, 66, 113, 118, 119–23,
  129
  camera of, 120–22, 125, 128
  Facebook acquisition of, 119–20
  founding of, 121
  misunderstanding of, 120
  named filters of, 120–22, 123
  revolutionary effect of, 122
International Standard Book Numbers
  (ISBNs), 90
Internet, 5–6, 34, 55–56, 62, 72, 102,
  178, 209–10
  ads on, 122
  aesthetics and morality of, 15
  anarchy of, 64
  as art, 7–8, 9–11, 13, 16, 51
  arts on, 49
  collaborative aspects of, 8, 21, 51
  connectivity of, 25–26, 94, 241
  cultural implications of, 15, 56
  dial-up access to, 136, 139
  early development of, 18–21

English usage on, 96–102
entrenchment of, 5, 21
epiphenomena of, 15
hierarchy of values on, 12, 18
hoaxes on, 11
influence of, 7
as integral part of humanity, 21
jargon of, 21
losses vs. gains of, 17–18, 177
magic of, 17–18, 241–42
poetry on, 57
qualities favored by, 10
response to critical methodologies
  by, 11
retreat from, 83
as seductive representation of the
  world, 7–8
surveillance with, 122
transformation of culture by, 21
two billion users of, 18
Internet of Things, 5
intuition, 28, 42, 210
Iowa-*Ploughshares* regime, 58
iPads, 5, 12, 17, 30, 88, 90, 91, 130–31,
  168
iPhone 5S, 113
iPhone 6 Plus, 113–14
iPhones, 5, 9, 12, 47, 90, 95, 113–14,
  131, 167–68, 174, 195, 210
  cameras on, 113–14, 119, 123, 128
  design of, 117, 118–19
  screens of, 113, 118–19
iPods, 178–83, 185–90, 193–95
Iran, anti-government movement in,
  13–14
Iraq, roadside bombs in, 200–201
Isaacson, Walter, 186, 187
Islam, 60–64
Isle of Wight, 58
Israel, 9
iTunes, 66, 147, 177, 183, 189, 195, 201
Ivory Coast, 61
"I Will Always Love You," 114

"Jackass" franchise, 35
James, Henry, 229

253

INDEX